HEALTH AND HUMAN DEVELOPMENT

ONCOLOGY

THE PROMISING FUTURE OF BIOMARKERS

Health and Human Development

Joav Merrick - Series Editor

National Institute of Child Health and Human Development, Ministry of Social Affairs, Jerusalem, Israel

Adolescent Behavior Research: International Perspectives
Joav Merrick and Hatim A Omar (Editors)
2007. ISBN: 1-60021-649-8

Complementary Medicine Systems: Comparison and Integration
Karl W Kratky
2008. ISBN: 978-1-60456-475-4

Pain in Children and Youth
P Schofield and J Merrick (Editors)
2008. ISBN: 978-1-60456-951-3

Alcohol-Related Cognitive Disorders: Research and Clinical Perspectives
Leo Sher, Isack Kandel and Joav Merrick (Editors)
2009. ISBN: 978-1-60741-730-9

Challenges in Adolescent Health: An Australian Perspective
David Bennett, Susan Towns, Elizabeth Elliott and Joav Merrick (Editors)
2009. ISBN: 978-1-60741-616-6

Children and Pain
P Schofield and J Merrick (Editors)
2009. ISBN: 978-1-60876-020-6

Living on the Edge: The Mythical, Spiritual, and Philosophical Roots of Social Marginality
Joseph Goodbread
2009. ISBN: 978-1-60741-162-8 (Hardcover)
2011. ISBN: 978-1-61470-192-7 (E-book)
2013. ISBN: 978-1-61122-986-8 (Softcover)

Obesity and Adolescence: A Public Health Concern
Hatim A Omar, Donald E Greydanus, Dilip R Patel and Joav Merrick (Editors)
2009. ISBN: 978-1-60692-821-9

Poverty and Children: A Public Health Concern
Alexis Lieberman and Joav Merrick (Editors)
2009. ISBN: 978-1-60741-140-6

Bone and Brain Metastases: Advances in Research and Treatment
Arjun Sahgal, Edward Chow and Joav Merrick (Editors)
2010. ISBN: 978-1-61668-365-8

Chance Action and Therapy: The Playful Way of Changing
Uri Wernik
2010. ISBN: 978-1-60876-393-1

Advances in Environmental Health Effects of Toxigenic Mold and Mycotoxins
Ebere Cyril Anyanwu
2011. ISBN: 978-1-60741-953-2

Alternative Medicine Yearbook 2009
Joav Merrick (Editor)
2011. ISBN: 978-1-61668-910-0

Behavioral Pediatrics, 3rd Edition
Donald E Greydanus, Dilip R Patel, Helen D Pratt and Joseph L Calles Jr. (Editors)
2011. ISBN: 978-1-60692-702-1

Chance Action and Therapy: The Playful Way of Changing
Uri Wernik
2011. ISBN: 978-1-61122-987-5

Child Health and Human Development Yearbook 2009
Joav Merrick (Editor)
2011. ISBN: 978-1-61668-912-4

International Aspects of Child Abuse and Neglect
Howard Dubowitz and Joav Merrick (Editors)
2010. ISBN: 978-1-61122-049-0

Advanced Cancer, Pain and Quality of Life
Edward Chow and Joav Merrick (Editors)
2011. ISBN: 978-1-61668-207-1

Climate Change and Rural Child Health
Erica Bell, Bastian M Seidel and Joav Merrick (Editors)
2011. ISBN: 978-1-61122-640-9

Clinical Aspects of Psychopharmacology in Childhood and Adolescence
DE Greydanus, JL Calles Jr., DP Patel, A Nazeer and J Merrick (Editors)
2011. ISBN: 978-1-61122-135-0

Drug Abuse in Hong Kong: Development and Evaluation of a Prevention Program
Daniel TL Shek, Rachel CF Sun and Joav Merrick (Editors)
2011. ISBN: 978-1-61324-491-3

Environment, Mood Disorders and Suicide
Teodor T Postolache and Joav Merrick (Editors)
2011. ISBN: 978-1-61668-505-8

International Aspects of Child Abuse and Neglect
Howard Dubowitz and Joav Merrick (Editors)
2011. ISBN: 978-1-60876-703-8

Narratives and Meanings of Migration
Julia Mirsky
2011. ISBN: 978-1-61761-103-2

Positive Youth Development: Evaluation and Future Directions in a Chinese Context
Daniel TL Shek, Hing Keung Ma and Joav Merrick (Editors)
2011. ISBN: 978-1-60876-830-1 (Hardcover)
2011. ISBN: 978-1-62100-175-1 (Softcover)

Positive Youth Development: Implementation of a Youth Program in a Chinese Context
Daniel TL Shek, Hing Keung Ma and Joav Merrick (Editors)
2011. ISBN: 978-1-61668-230-9

Principles of Holistic Psychiatry: A Textbook on Holistic Medicine for Mental Disorders
Soren Ventegodt and Joav Merrick
2011. ISBN: 978-1-61761-940-3

Public Health Yearbook 2009
Joav Merrick (Editor)
2011. ISBN: 978-1-61668-911-7

Randomized Clinical Trials and Placebo: Can You Trust the Drugs Are Working and Safe?
Søren Ventegodt and Joav Merrick
2011. ISBN: 978-1-61470-151-4 (E-book)
2012. ISBN: 978-1-61470-067-8 (Hardcover)

Rural Child Health: International Aspects
Erica Bell and Joav Merrick (Editors)
2011. ISBN: 978-1-60876-357-3

Rural Medical Education: Practical Strategies
Erica Bell and Craig Zimitat and Joav Merrick (Editors)
2011. ISBN: 978-1-61122-649-2

Self-Management and the Health Care Consumer
Peter William Harvey
2011. ISBN: 978-1-61761-796-6

Sexology from a Holistic Point of View
Soren Ventegodt and Joav Merrick
2011. ISBN: 978-1-61761-859-8

Social and Cultural Psychiatry Experience from the Caribbean Region
Hari D Maharajh and Joav Merrick (Editors)
2011. ISBN: 978-1-61668-506-5

The Dance of Sleeping and Eating among Adolescents: Normal and Pathological Perspectives
Yael Latzer and Orna Tzischinsky (Editors)
2011. ISBN: 978-1-61209-710-7

Understanding Eating Disorders: Integrating Culture, Psychology and Biology
Yael Latzer, Joav Merrick and Daniel Stein (Editors)
2011. ISBN: 978-1-61728-298-0 (Hardcover)
2011. ISBN: 978-1-61470-976-3 (Softcover)

Adolescence and Chronic Illness. A Public Health Concern
Hatim Omar, Donald E Greydanus, Dilip R Patel and Joav Merrick (Editors)
2012. ISBN: 978-1-60876-628-4

AIDS and Tuberculosis: Public Health Aspects
Daniel Chemtob and Joav Merrick (Editors)
2012. ISBN: 978-1-62081-382-9

Alternative Medicine Yearbook 2010
Joav Merrick (Editor)
2012. ISBN: 978-1-62100-132-4

Alternative Medicine Research Yearbook 2011
Joav Merrick (Editor)
2012. ISBN: 978-1-62081-476-5

Applied Public Health: Examining Multifaceted Social or Ecological Problems and Child Maltreatment
John R Lutzker and Joav Merrick (Editors)
2012. ISBN: 978-1-62081-356-0

Building Community Capacity: Minority and Immigrant Populations
Rosemary M Caron and Joav Merrick (Editors)
2012. ISBN: 978-1-62081-022-4

Building Community Capacity: Skills and Principles
Rosemary M Caron and Joav Merrick (Editors)
2012. ISBN 978-1-61209-331-4

Child and Adolescent Health Yearbook 2009
Joav Merrick (Editor)
2012. ISBN: 978-1-61668-913-1

Child and Adolescent Health Yearbook 2010
Joav Merrick (Editor)
2012. ISBN: 978-1-61209-788-6

Child Health and Human Development Yearbook 2010
Joav Merrick (Editor)
2012. ISBN: 8-1-61209-789-3

Conceptualizing Behavior in Health and Social Research: A Practical Guide to Data Analysis
Said Shahtahmasebi and Damon Berridge
2013. ISBN: 978-1-60876-383-2

Human Immunodeficiency Virus (HIV) Research: Social Science Aspects
Hugh Klein and Joav Merrick (Editors)
2012. ISBN: 978-1-62081-293-8

Our Search for Meaning in Life: Quality of Life Philosophy
Soren Ventegodt and Joav Merrick
2012. ISBN: 978-1-61470-494-2

Public Health Yearbook 2010
Joav Merrick (Editor)
2012. ISBN: 978-1-61209-971-2

Public Health Yearbook 2011
Joav Merrick (Editor)
2012. ISBN: 978-1-62081-433-8

Textbook on Evidence-Based Holistic Mind-Body Medicine: Basic Philosophy and Ethics of Traditional Hippocratic Medicine
Søren Ventegodt and Joav Merrick
ISBN: 978-1-62257-052-2

Textbook on Evidence-Based Holistic Mind-Body Medicine: Basic Principles of Healing in Traditional Hippocratic Medicine
Søren Ventegodt and Joav Merrick
2012. ISBN: 978-1-62257-094-2

Textbook on Evidence-Based Holistic Mind-Body Medicine: Healing the Mind in Traditional Hippocratic Medicine
Søren Ventegodt and Joav Merrick
2012. ISBN: 978-1-62257-112-3

Textbook on Evidence-Based Holistic Mind-Body Medicine: Holistic Practice of Traditional Hippocratic Medicine
Søren Ventegodt and Joav Merrick
2012. ISBN: 978-1-62257-105-5

Textbook on Evidence-Based Holistic Mind-Body Medicine: Research, Philosophy, Economy and Politics of Traditional Hippocratic Medicine
Søren Ventegodt and Joav Merrick
2012. ISBN: 978-1-62257-140-6

Textbook on Evidence-Based Holistic Mind-Body Medicine: Sexology and Traditional Hippocratic Medicine
Søren Ventegodt and Joav Merrick
2012. ISBN: 978-1-62257-130-7

The Astonishing Brain and Holistic Consciousness: Neuroscience and Vedanta Perspectives
Vinod D Deshmukh
2012. ISBN: 978-1-61324-295-7

Translational Research for Primary Healthcare
Erica Bell, Gert P Westert and Joav Merrick (Editors)
2012. ISBN: 978-1-61324-647-4

Treatment and Recovery of Eating Disorders
Daniel Stein and Yael Latzer (Editors)
2012. ISBN: 978-1-61470-259-7 (Hardcover)
2013. ISBN: 978-1-62808-248-7 (Softcover)

Adolescence and Sports
Dilip R Patel, Donald E Greydanus, Hatim Omar and Joav Merrick (Editors)
2013. ISBN: 978-1-60876-702-1

Advanced Cancer: Managing Symptoms and Quality of Life
Natalie Pulenzas, Breanne Lechner, Nemica Thavarajah, Edward Chow and Joav Merrick (Editors)
2013. ISBN: 978-1-62808-239-5

Alternative Medicine Research Yearbook 2012
Søren Ventegodt and Joav Merrick (Editors)
2013. ISBN: 978-1-62808-080-3

Building Community Capacity: Case Examples from Around the World
Rosemary M Caron and Joav Merrick (Editors)
2013. ISBN: 978-1-62417-175-8

Bullying: A Public Health Concern
Jorge C Srabstein and Joav Merrick (Editors)
2013. ISBN: 978-1-62618-564-7

Food, Nutrition and Eating Behavior
Joav Merrick and Sigal Israeli
2013. ISBN: 978-1-62948-233-0

Health and Happiness from Meaningful Work: Research in Quality of Working Life
Soren Ventegodt and Joav Merrick (Editors)
2013. ISBN: 978-1-60692-820-2

Health Promotion: Community Singing as a Vehicle to Promote Health
Joav Merrick (Editors)
2013. ISBN: 978-1-62618-908-9

Health Promotion: Strengthening Positive Health and Preventing Disease
Jing Sun, Nicholas Buys and Joav Merrick (Editors)
2013. ISBN: 978-1-62257-870-2

Health Risk Communication
Marijke Lemal and Joav Merrick(Editors)
2013. ISBN: 978-1-62257-544-2

Human Development: Biology from a Holistic Point of View
Søren Ventegodt, Tyge Dahl Hermansen and Joav Merrick (Editors)
2013. ISBN: 978-1-61470-441-6

Managed Care in a Public Setting
Richard Evan Steele
2013. ISBN: 978-1-62417-970-9

Mental Health from an International Perspective
Joav Merrick, Shoshana Aspler and Mohammed Morad (Editors)
2013. ISBN: 978-1-62948-519-5

Pain Management Yearbook 2011
Joav Merrick (Editor)
2013. ISBN: 978-1-62808-970-7

Pain Management Yearbook 2012
Joav Merrick (Editor)
2013. ISBN: 978-1-62808-973-8

Pediatric and Adolescent Sexuality and Gynecology: Principles for the Primary Care Clinician
Hatim A Omar, Donald E Greydanus, Artemis K Tsitsika, Dilip R Patel and Joav Merrick (Editors)
2013. ISBN: 978-1-60876-735-9

Public Health Concern: Smoking, Alcohol and Substance Use
Joav Merrick and Ariel Tenenbaum (Editors)
2013. ISBN: 978-1-62948-424-2

Public Health Yearbook 2012
Joav Merrick (Editor)
2013. ISBN: 978-1-62808-078-0

Alternative Medicine Research Yearbook 2013
Joav Merrick (Editor)
2014. ISBN: 978-1-63321-094-3

India: Health and Human Development Aspects
Joav Merrick (Editor)
2014. ISBN: 978-1-62948-784-7

Public Health: Improving Health via Inter-Professional Collaborations
Rosemary M. Caron and Joav Merrick (Editors)
2014. ISBN: 978-1-63321-569-6

Public Health Yearbook 2013
Joav Merrick (Editor)
2014. ISBN: 978-1-63321-095-0

Suicide from a Public Health Perspective
Said Shahtahmasebi and Joav Merrick
2014. ISBN: 978-1-62948-536-2

Alternative Medicine Research Yearbook 2014
Joav Merrick (Editor)
2015. ISBN: 978-1-63482-161-2

Cancer: Bone Metastases, CNS Metastases and Pathological Fractures
Breanne Lechner, Ronald Chow, Natalie Pulenzas, Marko Popovic, Na Zhang, Xiaojing Zhang, Edward Chow and Joav Merrick (Editors)
2015. ISBN: 978-1-63483-949-5

Cancer: Pain and Symptom Management
Breanne Lechner, Ronald Chow, Natalie Pulenzas, Marko Popovic, Na Zhang, Xiaojing Zhang, Edward Chow and Joav Merrick (Editors)
2015. ISBN: 978-1-63483-864-1

Cancer: Spinal Cord, Lung, Breast, Cervical, Prostate, Head and Neck Cancer
Breanne Lechner, Ronald Chow, Natalie Pulenzas, Marko Popovic, Na Zhang, Xiaojing Zhang, Edward Chow and Joav Merrick (Editors)
2015. ISBN: 978-1-63483-904-4

Cancer: Survival, Quality of Life and Ethical Implications
Breanne Lechner, Ronald Chow, Natalie Pulenzas, Marko Popovic, Na Zhang, Xiaojing Zhang, Edward Chow and Joav Merrick (Editors)
2015. ISBN: 978-1-63483-905-1

Forensic Psychiatry: A Public Health Perspective
Leo Sher and Joav Merrick (Editors)
2015. ISBN: 978-1-63483-339-4

Leadership and Service Learning Education: Holistic Development for Chinese University Students
Daniel TL Shek, Florence KY Wu and Joav Merrick (Editors)
2015. ISBN: 978-1-63483-340-0

Mental and Holistic Health: Some International Perspectives
Joseph L. Calles Jr., Donald E Greydanus and Joav Merrick (Editors)
2015. ISBN: 978-1-63483-589-3

Pain Management Yearbook 2014
Joav Merrick (Editor)
2015. ISBN: 978-1-63482-164-3

Public Health Yearbook 2014
Joav Merrick (Editor)
2015. ISBN: 978-1-63482-165-0

Cancer: Treatment, Decision Making and Quality of Life
Breanne Lechner, Ronald Chow, Natalie Pulenzas, Marko Popovic, Na Zhang, Xiaojing Zhang, Edward Chow and Joav Merrick (Editors)
2015. ISBN: 978-1-63483-863-4

Alternative Medicine Research Yearbook 2015
Joav Merrick (Editor)
2016. ISBN: 978-1-63484-511-3

Public Health Yearbook 2015
Joav Merrick (Editor)
2016. ISBN: 978-1-63484-514-4

Quality, Mobility and Globalization in the Higher Education System: A Comparative Look at the Challenges of Academic Teaching
Nitza Davidovitch, Zehavit Gross, Yuri Ribakov, and Anna Slobodianiuk
2016. ISBN: 978-1-63484-986-9

Cannabis: Medical Aspects
Blair Henry, Arnav Agarwal, Edward Chow, Hatim A. Omar, and Joav Merrick
2017. ISBN: 978-1-53610-510-0

Palliative Care: Psychosocial and Ethical Considerations
Blair Henry, Arnav Agarwal, Edward Chow, and Joav Merrick
2017. ISBN: 978-1-53610-607-7

Oncology: The Promising Future of Biomarkers
Anthony Furfari, George S. Charames, Rachel McDonald, Leigha Rowbottom, Azar Azad, Stephanie Chan, Bo Angela Wan, Ronald Chow, Carlo DeAngelis, Pearl Zaki, Edward Chow, Joav Merrick (Editors)
2017. ISBN: 978-1-53610-608-4

HEALTH AND HUMAN DEVELOPMENT

ONCOLOGY

THE PROMISING FUTURE OF BIOMARKERS

ANTHONY FURFARI,
GEORGE S CHARAMES,
RACHEL MCDONALD,
LEIGHA ROWBOTTOM,
AZAR AZAD,
STEPHANIE CHAN,
BO ANGELA WAN,
RONALD CHOW,
CARLO DEANGELIS,
PEARL ZAKI,
EDWARD CHOW AND
JOAV MERRICK
EDITORS

New York

Library of Congress Cataloging-in-Publication Data

ISBN: 978-1-53610-608-4

Library of Congress Control Number: 2016959485

Published by Nova Science Publishers, Inc. † New York

CONTENTS

INTRODUCTION

In: Oncology
Editors: A. Furfari et al.

ISBN: 978-1-53610-608-4

Chapter 1

BIOMARKERS AND ONCOLOGY

Anthony Furfari*[1]*, MD(C),
George S Charames*[2,3,4,*]*, PhD,
Leigha Rowbottom*[1]*, MD(C), Rachel McDonald*[1]*, MD(C),
Carlo DeAngelis*[1]*, PharmD, Azar Azad*[2,3,5]*, PhD,
Stephanie Chan*[1]*, BSc(C), Bo Angela Wan*[1]*, MPhil,
Ronald Chow*[1]*, BMSc(C), Pearl Zaki*[1]*, BSc(C),
Edward Chow*[1]*, MBBS and
Joav Merrick*[6-10]*, MD, MMedSc, DMSc

[1]Odette Cancer Centre, Sunnybrook Health Sciences Centre, University of Toronto, Toronto, Ontario, Canada
[2]Pathology and Laboratory Medicine, Mount Sinai Hospital, Toronto, Ontario, Canada
[3]Laboratory Medicine and Pathobiology, University of Toronto, Toronto, Ontario, Canada
[4]Lunenfeld-Tanenbaum Research Institute, Sinai Health System, Toronto, Ontario, Canada
[5]Mount Sinai Services Inc., Toronto, Ontario, Canada
[6]National Institute of Child Health and Human Development, Jerusalem, Israel

[*] Correspondence: George S Charames PhD, FACMG, Pathology and Laboratory Medicine, Mount Sinai Hospital, Suite 6-500, 600 University Ave, Toronto, Ontario, M5G 1X5, Canada. E-mail: George.Charames@sinaihealthsystem.ca.

[7]Office of the Medical Director, Health Services, Division for Intellectual and Developmental Disabilities, Ministry of Social Affairs and Social Services, Jerusalem, Israel
[8]Division of Pediatrics, Hadassah Hebrew University Medical Center, Mt Scopus Campus, Jerusalem, Israel
[9]Kentucky Children's Hospital, University of Kentucky School of Medicine, Lexington, Kentucky, United States
[10]Center for Healthy Development, School of Public Health, Georgia State University, Atlanta, United States of America

INTRODUCTION

Cancer is a term that describes a constellation of diseases that affect a large proportion of the global population. By 2020, it is estimated there will be 15 million new cancer cases worldwide and a corresponding 12 million cancer-related deaths (1). There are many efficacious treatments for cancer with both curative and palliative intents, which include surgery, chemotherapy, and radiotherapy. Regardless of efficacy, such treatments can result in distressing symptoms, including pain, fatigue, nausea, vomiting, and reduced function. Compounded with very similar symptoms from the disease itself, toxicities can ultimately lead to diminished quality of life (QOL). Not only is there a significant patient impact, but the expense of cancer care comprises a significant proportion of health care spending. With projected increases in the incidence of cancer, new healthcare strategies are being explored to provide more efficacious treatments with minimal toxicity in order to deliver patient centered care while ensuring healthcare remains a feasible expenditure.

In recent years, the use of biomarkers in cancer research and practice has become a widely discussed topic of exploration for researchers and clinicians alike. A biomarker is a biological property that can be measured and used as a tool in the prevention, diagnosis, and treatment of specific illnesses, including cancer (2). Within the literature, biomarkers are seen as an opportunity to reap large rewards from both an economic and clinical standpoint. The excitement surrounding biomarker research and development was described in Andrea Renaldi's article *"Teaming up for biomarker future"* (3). The article highlighted the importance of biomarkers of disease state as a main promoter of proteomic research, and goes on to quote professionals in the field expressing their thoughts and concerns on the topic. With that established, the main theme of the article seemed to be a "call to arms" in realizing the future of biomarkers in the clinical environment, and acts as a catalyst to realize that much work needs done in order to efficiently use biomarkers in clinical practice. The future of biomarkers, put forth by Andrea Renaldi, is in the collaboration and sharing of biomarker research, which acts to further develop our knowledge of biomarkers allowing the field to move forward (3).

TREATMENT DECISION

One area of particular interest regarding biomarkers is the potential they have to improve treatment decisions. For instance, physicians may be able to use genetic biomarkers in order to properly predict a patient's response to drug therapies and from such screen, prescribe a more personalized course of treatment. This potential holds true across both systemic therapy, including chemotherapy, and more targeted pharmaceuticals, such as analgesics. One example of the use of biomarkers in practice is the prescription of targeted therapy (erlotinib, gefitinib, and afatinib) for non-small cell lung cancer patients who are mutation positive for epidermal growth factor receptor (EGFR). A systematic review by Greenhaulgh et al. (4) concluded that such targeted treatments are associated with fewer toxicities and improved progression-free survival when compared to standard chemotherapy.

Beyond anti-cancer medications, there is potential for biomarkers to be utilized in the treatment of cancer-related pain. Previous research has noted considerable differences amongst various patient groups in the metabolism of analgesics, including opioids, steroids, narcotics, and anti-inflammatory medications. For instance, Kirchheiner et al. (5) examined the gene encoding enzyme CYP2D6, a morphine metabolizer, and found it had undergone a single polymorphic change in specific ethnic groups. Polymorphisms in this gene resulted in a 150% greater response to morphine in Caucasian patients indicating that drug efficacy was significantly higher and that caution should be used when prescribing in order to avoid overdose (5).

As discussed, patients affected by cancer may experience a variety both of cancer- and treatment-related symptoms, including pain, pain flare, fatigue, nausea, vomiting, psychosocial issues, and many others. In addition to treating these symptoms, predicting them through biomarkers is advantageous as such foresight may instigate the initiation of strategies to prevent likely symptoms. One instance where biomarkers have been utilized in this setting is C-reactive protein, which has been shown to predict fatigue in breast cancer patients prior to receiving chemotherapy (6). Following the identification of such a biomarker, healthcare providers may recommend identified patients to commence exercise programs to proactively reduce fatigue.

Prognosis

An additional factor in guiding treatment and care decisions is prognosis. Various prognostic models have been developed in the cancer setting which incorporate different clinical and disease characteristics, including patient performance status and extent of visceral metastases. Biomarkers have the potential to act as objective prognostic indicators which may improve the accuracy of these estimations. With improved ability to estimate survival, physicians can better inform their patients of their disease status, better prescribe treatments that will be useful given the disease trajectory, and better discuss end-of-life concerns.

With technological advances, we have amassed numerous ways in which biomarkers are able to be collected; for instance, through blood, saliva, urine, tissue samples, and almost any bodily component that contains DNA or RNA. This makes finding biomarkers an extremely viable option for dealing with patients afflicted with a variety of illnesses. This is because it allows for simple extraction methods, such as urine samples, as well as more invasive extraction methods, such as blood samples or biopsies, to both yield results with high utility. Moreover, this helps in the exclusion of guesswork, and may ultimately lead to a more comprehensive diagnosis of the situation. Biomarkers have the potential to become a staple in precision medicine which incorporates genetics and discrete individual characteristics (i.e., environment and lifestyle) to allow for strategic treatment options.

This book synthesizes existing literature biomarker research that has been conducted in relation to the oncological field, and the potential effect it may have on the future treatment of cancer patients. In these chapters, we will delve into the literature regarding drug metabolism, symptom burden, radiation response, susceptibility to radiation related toxicities, and prognosis within the oncology field.

ACKNOWLEDGMENTS

We thank the generous support of Joey and Mary Furfari Cancer Research Fund.

REFERENCES

[1] Schottenfeld D, Beebe-Dimmer J. Alleviating the burden of cancer: a perspective on advances, challenges, and future directions. Cancer Epidemiol Biomarkers Prev 2006;15(11):2049-55.

[2] Fuentes-Arderiu X. What is a biomarker? It's time for a renewed definition. Clin Chem Lab Med 2013;51(9):1689-90.

[3] Renaldi A. Teaming up for biomarker future. Many problems still hinder the use of biomarkers in clinical practice, but new public-private partnerships could improve the situation. EMBO Rep 2011;12(6):500-4.

[4] Greenhalgh J, Dwan K, Boland A, Bates V, Vecchio F, Dundar Y, et al. First-line treatment of advanced epidermal growth factor receptor (EGFR) mutation positive non-squamous non-small cell lung cancer. Cochrane Database Syst Rev 2016;(5):CD010383.

[5] Kirchheiner J, Schmidt H, Tzvetkov M, Keulen J, Lötsch J, Roots I, et al. Pharmacokinetics of codeine and its metabolite morphine in ultra-rapid metabolizers due to CYP2D6 duplication. P Pharmacogenomics J 2007;7(4):257-65.

[6] Pertl MM, Hevey D, Boyle NT, Hughes MM, Collier S, O'Dwyer AM, et al. C-reactive protein predicts fatigue independently of depression in breast cancer patients prior to chemotherapy. Brain Behav Immun 2013; 34:108-19.

SECTION ONE: BIOMARKERS

In: Oncology
Editors: A. Furfari et al.

ISBN: 978-1-53610-608-4

Chapter 2

GENETIC VARIANTS AND BIOLOGICAL MARKERS OF CANCER-RELATED PAIN SENSITIVITY

Rachel McDonald*[1]*, MD(C), Adam Bobrowski*[1]*, BSc, Matthew Choi*[1]*, BSc, Stephanie Chan*[1]*, BSc(C), Pearl Zaki*[1]*, BSc(C), Bo Angela Wan*[1]*, MPhil, Anthony Furfari*[1]*, MD(C), Leigha Rowbottom*[1]*, MD(C), Henry Lam*[1]*, MLS, Ronald Chow*[1]*, BMSc(C), George S Charames*[2,3,4]*, PhD, Azar Azad*[2,3,5]*, PhD, Edward Chow*[1]*, MBBS and Carlo DeAngelis*[1,*]*, PharmD

[1]Odette Cancer Centre, Sunnybrook Health Sciences Centre, University of Toronto, Toronto, Ontario, Canada
[2]Mount Sinai Services Inc., Toronto, Ontario, Canada
[3]Pathology and Laboratory Medicine, Mount Sinai Hospital, Toronto, Ontario, Canada
[4]Lunenfeld-Tanenbaum Research Institute, Sinai Health System, Toronto, Ontario, Canada
[5]Laboratory Medicine and Pathobiology, University of Toronto, Toronto, Ontario, Canada

[*] Correspondence: Carlo DeAngelis, PharmD, Odette Cancer Centre, Sunnybrook Health Sciences Centre, 2500 Bayview Ave, Toronto, Ontario, M4N 3M5, Canada. E-mail: carlodeangelis@sunnybrook.ca.

Pain is one of the most commonly reported and least understood symptoms amongst cancer patients and survivors. Improper management of pain can lead to significant reductions in physical and psychological wellbeing, and diminished quality of life. Patients with similar types and stages of cancer can present vastly different accounts of severity of pain, causes of pain, and responses to treatment methods aimed to reduce pain such as radiation therapy and analgesics. These differences suggest an underlying genetic etiology, and highlight the lack of understanding of how cancer pain develops. This chapter summarizes 25 studies on biomarkers of pain in cancer patients. Many of these studies show that up-regulation of inflammatory biomarkers and hormones is associated with increased reports of pain and pain severity. In addition, single nucleotide polymorphisms within common inflammatory genes and signaling protein genes have been associated with risk for cancer pain. The chapter concludes that there are a variety of biomarkers with clinical utility in predicting cancer pain, and may also act as potential therapeutic targets. Future studies should look to examine their role in the progression and development of pain, and should also look for biomarkers that can distinguish between the different types of cancer pain.

INTRODUCTION

Pain is one of the most commonly experienced and distressing symptoms reported by cancer patients in both their disease and therapeutic trajectories (1). Cancer patients have reported a greater fear of disease-related pain and suffering than death (2). As many as 90% of cancer patients experience pain during the disease trajectory (1), with 59% experiencing pain during therapy and 65% during advanced disease (2). Additionally, 60-80% of cancer patients experience moderate to severe pain (3). Pain can be experienced regardless of primary cancer site and, if unrelieved, can lead to significant functional, emotional, spiritual and psychosocial impairments, ultimately resulting in diminished quality of life (QOL). Despite treatment guidelines set by the World Health Organization (WHO), 52-77% of patients suffer from cancer pain and the pain symptoms of up to 45% of patients are mismanaged and inadequately controlled (1).

Cancer pain can arise from a variety of mechanisms. Neuropathic pain is caused by damage to the central and peripheral nervous systems, while nociceptive pain is a result of primary tumours or metastatic processes activating somatic or visceral nociceptors. The encroachment of a tumour on adjacent tissue, tissue damage, and blood vessel obstruction also causes pain (4). These differences in etiology alter the treatment target and management of pain.

Treatment side effects further complicate the therapeutic process by potentially causing cancer patients additional pain. For instance, breast cancer patients with bony metastases receiving aromatase inhibitors can experience pain directly related to the bone metastases or from treatment-induced myalgia and arthralgia (5). Similarly, approximately one third of patients undergoing palliative radiotherapy for painful bone metastases can experience a transient exacerbation of pain, known as pain flare (6). Although this is a well-described phenomenon, it may be difficult to differentiate pain flare from disease progression at its initial onset. Furthermore, the different types of pain-such as neuropathic or nociceptive-each require a different management strategy. Thus, proper assessment of the characteristics of pain is required to ensure the most effective pain management practices are followed.

One of the difficulties in assessing and managing pain symptoms is the genetic variation contributing to differences in nociceptive and analgesic response among individuals. Animal studies investigating the existence of heritable differences in pain sensitivity reported that genomic variations contribute anywhere from 28-76% of the variability in pain experience (7). The varying severities of pain from patients suffering from similar types and stages of cancer suggest that genetic variability within humans produces similar differences in humans.

Given the current lack of adequate palliation of pain and the rising incidence of cancer in the global population, it is critical that further research is conducted to better elucidate variations in pain sensitivity. Specifically, identifying genetic variants or other biological biomarkers that could predict an individual's response to painful stimuli could better inform physicians and regulatory bodies regarding appropriate treatment protocols. The purpose of this chapter is to elucidate the biological markers, including single nucleotide polymorphisms (SNPs), related to cancer-related pain sensitivity.

OUR LITERATURE SEARCH

A literature search was conducted in Ovid MEDLINE (R) spanning 1946 to June Week 1 2016 and Embase Classic + Embase spanning 1947 to 2016 Week 24. The search was limited to English articles and restricted to human studies only. Key search terms used included "neoplasm," "cancer pain," "inflammation," "biological marker," "allele," and "genomics."

Results generated from the search were screened first by title and abstract by two authors (RM, AB) based on pre-determined inclusion and exclusion criteria. For all remaining articles, full-text was retrieved and thoroughly screened until a final list of articles was determined. If there were discrepancies amongst the authors, discussion ensued until a consensus on inclusion or exclusion was achieved. To be included in the review, articles were required to discuss specific genomic variations or biological biomarkers in the context of cancer pain. Articles discussing pain directly related to analgesics, chemotherapy, radiotherapy, or bisphosphonate use were excluded. Additionally, any review articles, case studies, or those conducted in animal models were excluded. Figure 1 illustrates the selection process.

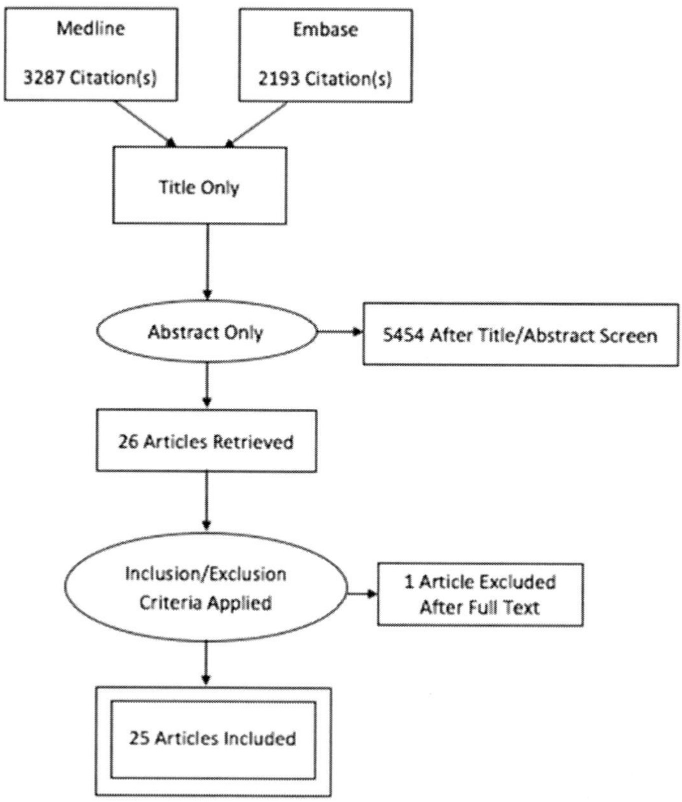

Figure 1. Literature review screening process.

Data of interest were extracted from the final list of selected articles independently by two authors (RM, AB). These included information regarding primary cancer site, number of patients, cause of pain, and genomic variants and other biological biomarkers investigated for altering pain sensitivity. The data extracted by both authors were compared to ensure accuracy; if any discrepancies were found the article was reviewed by both authors and a consensus was achieved.

OUR FINDINGS

A review of Embase and Medline yielded a total of 5,480 articles, of which 25 articles were deemed to be eligible for inclusion following title and abstract review, and full-text screening (see Figure 1). These studies reported on SNPs and biological markers associated with cancer-related pain sensitivity. These symptom-related biological markers are elaborated below and summarized in Table 1.

Breast cancer

Using a sample population of 322 breast cancer survivors, Lengacher et al. found that levels of IL6 were significantly associated with pain (p=0.02) and QOL (p=0.05) (8). They also found that levels of cortisol were significantly associated with physical health (p=0.05) (8).

McCann et al. described genetic factors influencing occurrence rate and characteristics of preoperative breast cancer pain (9). 390 breast cancer patients who completed self-administered questionnaires were genotyped for 103 SNPs in 15 candidate genes involved in the inflammatory response. Over a quarter of women included in the study (28.2%) suffered from pre-surgical breast pain. Of this sample, a greater proportion of patients experiencing pain were of a non-white ethnicity (p=0.018). Carriers of the minor allele (CT and TT) for a SNP rs2110726 in *IL1R1* had a 53% lower likelihood of reporting breast pain prior to surgery (p=0.007). The minor allele of *IL13* SNP rs1295686 was associated with a 1.57-fold greater risk of experiencing breast cancer pain prior to surgery (p=0.019) (9).

Table 1. Summary of literature findings

Author (Reference)	Basics	Cancer Type	Gene or Gene Product	Gene Variability	Result	Procedure Used
Lengacher et al. 2015 (8)	Mindfulness-based stress reduction for breast cancer, evaluate it in reducing cortisol and IL6 Salivary cortisol and IL6 assessed baseline + 6 weeks n=322	Breast Cancer (survivors)	*IL6* *Cortisol*	N/A	Relationship between IL6 and pain (r=0.21, p=0.02) at baseline.	Salivary
Stephens et al. 2014 (11)	Persistent pain following breast cancer surgery Pain ratings – Breast Symptoms Questionnaire and Postsurgical	Breast cancer	*IL1R2* *IL10*	rs11674595 Haplotype A8 → composed of 7SNPs: rs3024505 C allele, rs3024498 G allele, rs3024496 C allele, rs1878672 G allele, rs1518111 A allele, rs1518110	Homozygous for *C* allele associated with 36.1-fold increase in odds of belonging to severe breast pain class	Blood

Author (Reference)	Basics	Cancer Type	Gene or Gene Product	Gene Variability	Result	Procedure Used
	Pain Questionnaire n=126 (no breast pain) n=46 (breast pain)			T allele, rs3024491 T allele	Each dose of haplotype A8 decreased odds of belonging to severe breast pain class by 79%	
McCann et al. 2012 (9)	Analyzed biomarkers of breast pain in women prior to breast ca surgery (n=302)	Breast cancer	*IL1R1* *IL13*	rs2110726 (p=0.007) rs1295686 (p=0.019)	Carriers of minor allele (*CT* + *TT*) had 53% decrease in the odds of reporting breast pain prior to surgery (p=0.007) Each dose of the minor allele was associated with a 1.57 fold increase in the odds of reporting breast pain prior to surgery	Blood

Table 1. (Continued)

Author (Reference)	Basics	Cancer Type	Gene or Gene Product	Gene Variability	Result	Procedure Used
Liu et al. 2012 (26)	Evaluated whether *TLC1A* SNPs altered influence cytokine expression and explore mechanism of *TLC1A*-related aromatase inhibitor induced side effects	Breast cancer	*TLC1A*	rs7158782 rs7159713 rs2369409 rs11849538 (lowest p value) and forms a functional estrogen response element All on chromosome 14	Estrogen induced *TCL1A* expression was significantly greater in lymphoblastoid cell line carrying variants instead of WT *TLC1A, IL17RA, IL17, IL1R2, IL12RB2,* and *IL12* all altered by increased concentrations of estradiol *TLC1A* and *IL1RA* expression increased in variant lines from exposure to estradiol. *IL17, IL1R2, IL12RB2,*	Collection of human lymphoblastoid cell lines and genetic analysis

Author (Reference)	Basics	Cancer Type	Gene or Gene Product	Gene Variability	Result	Procedure Used
					and *IL12* expression increased in WT lines. NFκB activity rose over 3x with estradiol antagonist (mimicking estrogen withdrawal) in variant lines, while wild-type only decreased slightly. This may be contributing to increased pain.	
Langford et al. 2014 (10)	Potassium channel gene variants have been associated with perioperative breast cancer pain. SNPs and haplotypes of	Breast cancer	*KCNS1* *KCNJ3* *KCNJ6* *KCNK9*	*KCNS1*: rs4499491 *KCNJ3*: rs7574878 and haplotype E1 (rs2591168 and rs2591172) *KCNJ6*: rs2835914, rs8129919, rs2836050	Of the 10 genes studied, 7 SNPs and 1 haplotype among 4 candidate genes were associated with pain.	Patients were administered the Karnofsky Performance Status (KPS) questionnaire to evaluate

Table 1. (Continued)

Author (Reference)	Basics	Cancer Type	Gene or Gene Product	Gene Variability	Result	Procedure Used
	10 potassium transporter genes were evaluated.			*KCNK9*: rs3780039, rs11166921		functional status, the Self-Administered Co-Morbidity Questionnaire (SCQ) and a pain assessment questionnaire.
					KCNS1 rs4499491 homozygous for A allele had a 3-fold increase in odds of reporting preoperative breast cancer pain. *KCNJ3* rs7574878 heterozygous/ homozygous for *G* allele had a 48% reduction in preoperative breast pain	Blood samples were processed using the PUREGene DNA Isolation and the Golden Gate System Platform.

Author (Reference)	Basics	Cancer Type	Gene or Gene Product	Gene Variability	Result	Procedure Used
					Haplotype E1 of the *KCNJ3* gene (*A* allele at rs2591168 and *G* allele at rs2591172 was linked to a 1.7-fold increase in the odds of reporting perioperative pain. *KCNJ6* rs2835914 homozygous/ heterozygous for *C* allele had a 52% reduction in odds of reporting perioperative breast pain. Each dose of the *A* allele in *KCNJ6* rs8129919 leads to a 2.1-fold increase in odds of reporting perioperative breast pain.	

Table 1. (Continued)

Author (Reference)	Basics	Cancer Type	Gene or Gene Product	Gene Variability	Result	Procedure Used
					KCNJ6 rs2836050 homozygous for *T* allele had a 3.6-fold increase in odds of perioperative breast pain. *KCNK9* rs3780039 heterozygous/ homozygous for *G* allele had 1.9-fold increase in odds of reporting perioperative breast pain. *KCNK9* rs11166921 homozygotes of the *A* allele had a 2.4-fold increase odds of reporting preoperative breast pain.	

Author (Reference)	Basics	Cancer Type	Gene or Gene Product	Gene Variability	Result	Procedure Used
Tagawa et al. 2015 (19)	Secondary analysis of TAXYNERGY Collected 2 sets of circulating tumor cell counts (CTCs) – screening and baseline, w no intervening treatment 63 men randomized; evaluable in 62	Prostate, chemo-naive	CTC	N/A	Higher CTC counts associated w higher baseline LDH and pain assessments Analgesic score + present pain intensity score remained sig after multivariate analysis	Blood
Reyes-Gibby et al. 2007 (12)	Explored associations between SNPs in 3 cytokine genes, and pain severity in newly diagnosed NSCLC patients (total n=695; n=446 whites,	Lung cancer	*TNF* *IL6* *IL8*	*TNF*: 308 G/A *IL6*: -174G/C *IL8*: -251T/A	*IL8* (*TT*, 13%; *TA* + *AA*, 87%; p=0.04) significantly associated with severe pain among white patients	Blood

Table 1. (Continued)

Author (Reference)	Basics	Cancer Type	Gene or Gene Product	Gene Variability	Result	Procedure Used
	125 African-Americans, 35 Hispanics)				In multivariate analysis: variant alleles in *IL8 -251T/A* (OR=2.35; 95% CI=1.10-5.03; p=0.03) persisted as a significant factor for severe pain for white patients	
Rausch et al. 2012 (17)	n=1149 Caucasian lung ca patient survivors Early survivors <3yrs Middle term survivors 3-5 yrs Long term survivors >5 yrs	Lung cancer	*PTGS2* *LTA*	*PTGS2*: rs5277 and rs5275 *LTA*: rs1799964	*PTGS2:* 3-5 year survivors, associated with pain; carriers of one or two minor (*G*) alleles reported higher scores for pain, lower social functioning, and lower mental health.	Blood serum

Author (Reference)	Basics	Cancer Type	Gene or Gene Product	Gene Variability	Result	Procedure Used
					LTA: 5+ yr survivors, associated with pain; people carrying one or two minor (*G*) alleles reported lower pain	
Reyes-Gibby et al. 2013 (15)	Examined a host of immune response genes that underlie the co-occurrence of severe pain, depression and fatigue to identify patients in the severe or non-severe symptom clusters (n=599)	Lung cancer	*ENOS* *IL1B* *TNFR2* *PTGS2* *IL10RB*	*ENOS: -1474 T/A* *IL1B: T-31C* *TNFR2*: Met1^{96}Arg *PTGS2: exon 10+837T>C* *IL10RB: Lys47Glu*	Of the 55 SNPs reviewed, the mutant alleles were additive in effect and predictive for symptom clusters. Non-genetically, stage of disease and sex were strongly predictive of high symptom burden.	Patients rated pain on 1— point scale; fatigue was assessed using items from the SF- Patients charts were used to resolve stage of disease, medications and comorbidity history. DNA extracted from lymphocytes and SNP s were genotyped

Table 1. (Continued)

Author (Reference)	Basics	Cancer Type	Gene or Gene Product	Gene Variability	Result	Procedure Used
						using SNPlex Bayesian approach used to consider candidate variables as being significant.
Reyes-Gibby et al. 2013 (16)	Examined variants of 37 inflammation genes for their role as biomarkers of risk for severe pain, depressed mood, and fatigue in non-Hispanic white patients with non-small cell lung cancer	Lung cancer	*IL8* *IL10RB* *IL1A*	*IL8:T251A* *IL10RB: Lys47Glu* *IL1A: C-889T, CT or TT*	*IL8* was the most relevant genetic factor for pain, depressed mood, and fatigue amongst patients with advanced-stage disease Women with genotypes *Lys_Glu* or *Glu_Glu* in *IL10RB Lys47Glu* polymorphism had	Pain, fatigue, and depressed mood assessed before cancer treatment. Blood samples drawn and genetically analyzed for SNPs in inflammatory genes associated with pain, depressed mood, and fatigue.

Author (Reference)	Basics	Cancer Type	Gene or Gene Product	Gene Variability	Result	Procedure Used
					0.49x lower risk of severe fatigue compared with *Lys_Lys* genotype *IL1A:* Male patients with *IL1A C-889T, CT* or *TT* genotypes had lower risk of fatigue compared with *CC* genotype.	
Reyes-Gibby et al. 2009 (14)	Evaluated 59 SNPs of inflammatory biomarkers in newly diagnosed lung cacner patients (n=667, Caucasian patients) and associated the SNPs to pain severity	Lung cancer	*PTGS2* *TNF* *NFKB*	*PTGS2:* exon *10+837T>C* *TNF: 308GA* rs1800629 *NFKB: Ex6+50C>T* rs5904	Patients with *CC* genotypes for *PTGS2* exon *10+837T>C* (rs5257) had 5 % lower risk for severe pain relative to carriers of *TT* and *TC* genotypes Additive model for *TNF-308GA* (rs1800629) and *NFKBIA Ex6+50C>T*	Blood

Table 1. (Continued)

Author (Reference)	Basics	Cancer Type	Gene or Gene Product	Gene Variability	Result	Procedure Used
					(rs5904) predictive of severe pain In multigene model for severe pain, only *TNF-308GA* sig predicted severe pain, other 2 were borderline Each protective allele decreased risk for severe pain by as much as 38%, even after adjustment for non-genetic variables	
Reyes-Gibby et al. 2008 (13)	To examine polymorphisms within previously studied cytokine genes for their role in	Lung cancer	*TNF* *IL6* *IL8*	*308G/A* *174G/C* *251T/A*	*TNF* polymorphism were significantly associated with pain severity at first follow-up visit (p=0.04)	Pain questionnaires to assess pain relief using the Edmonton Symptom

Author (Reference)	Basics	Cancer Type	Gene or Gene Product	Gene Variability	Result	Procedure Used
	explaining variability in pain and analgesic response for patients who were referred for pain treatment n=140				Stratified by sex: significant differences in pain severity by *TNF* genotypes for men and by *IL6* genotypes for women ($p<0.05$) Only polymorphisms in *TNF* remained as a significant factor for pain severity ($b=0.226$; $p=0.036$) after multivariate analysis Improvement in pain scores was observed for carriers of polymorphisms in *IL-6 −174GC* and *IL-8 −251T/A*	Assessment System. Blood samples were taken for genetic analysis

Table 1. (Continued)

Author (Reference)	Basics	Cancer Type	Gene or Gene Product	Gene Variability	Result	Procedure Used
					TNF-308GA: *AA* genotype exhibited no improvement in pain scores relative to the *GG* and *GA* genotypes	
Wang et al. 2010 (18)	n=62 undergoing chemoradiation → platinum taxane based chemo regiment during radiotherapy	Lung cancer (non-small cell)	IL6, IL8, IL10, IL12p40p70, IL1RA, TNFα, sTNF-R1	N/A	Increase in IL6 associated with increase in mean severity of 5 most severe symptoms (pain, fatigue, disturbed sleep, lack of appetite, sore throat). During CXRT, serum sTNFR1 was only cytokine whose increased was sig associated w increase in 15 symptom component score	Blood samples prior to start of CXRT, weekly thereafter for approximately 8 weeks of therapy

Author (Reference)	Basics	Cancer Type	Gene or Gene Product	Gene Variability	Result	Procedure Used
Nguyen et al. 2012 (25)	Compared symptom burden of individuals with CML and systemic mastocytosis (SM) to those with classic MPNs n=495 patients	Myeloproliferative neoplasms	*JAK2*	*V617F*	Significantly correlated with bone pain (p=0.0341) – but not significant when all 6 symptoms analyzed together	Saliva
Steel et al. 2010 (21)	Determined the prevalence of several symptoms including pain in 206 hepatobiliary carcinoma (HBC) patients and determined if symptom clusters were associated with any biomarkers	Liver cancer	Eosinophils	NA	Symptomatic cluster had significantly higher eosinophils at three months (3.1 vs. 2.9 and 1.6) and six months (3.6 vs. 3.0 and 1.5) compared to asymptomatic and fatigued symptom cluster, respectively	Blood

Table 1. (Continued)

Author (Reference)	Basics	Cancer Type	Gene or Gene Product	Gene Variability	Result	Procedure Used
					Association between FACT-Pain and eosinophils: $z=1.914$, $p=0.056$. Eosinophils and pain significantly associated with depressive symptoms and eosinophils	
Engler et al. 1998 (27)	Examined the effect of bisphosphonate therapy on bone resorption parameters in cancer patients with malignant osteolytic bone disease, the diagnostic efficacy of bone	Advanced malignant osteolytic bone disease	N-terminal telopeptide collagen type I (NTx) Pyridinoline (PYD) Deoxypyridinoline (DPD)	N/A	Two bone resorption markers were significantly predictive for pain intensity reduction during bisphosphonate therapy: urinary free DPD excretion rate, and serum ICTP levels.	HPLC measurement of PYD and DPD, acid hydrolysis of urine sample, PYD extraction and separation, ELISA with urine sample

Author (Reference)	Basics	Cancer Type	Gene or Gene Product	Gene Variability	Result	Procedure Used
	matrix markers for the presence of skeletal metastases in patients with carcinomas, predictive ability of progression of skeletal tumour involvement under bisphosphonate therapy, predictive ability of reduction in pain intensity under bisphosphonate therapy, and predictive ability of increase in bone mineral density during bisphosphonate therapy.		C-terminal telopeptide of collagen type I (CTx)		Serum alkaline phosphatase activity also showed significant predictive value. NTx did not reach statistical significance, whereas urinary calcium excretion rate or CrossLaps excretion rate showed no effect at all. Bone AP did not show statistically significant predictive value.	

Table 1. (Continued)

Author (Reference)	Basics	Cancer Type	Gene or Gene Product	Gene Variability	Result	Procedure Used
Reyes-Gibby et al. 2009 (20)	Analyzed 13 polymorphisms in inflammatory cytokines and correlated to pain severity in 156 white patients	Pancreatic cancer	*IL1b, IL6, IL8, IL10, IL18, TNF, NFKB1*	*IL8-251T/A* p<0.026) rs4073 *IL1 B-31T/C* (p<0.044)	Significantly correlated with severe pain on univariate analysis Polymorphism in *IL8* was most important predictor of severe pain, followed by sex	Blood
Albulescu et al. 2013 (24)	Analyze serum pro- and anti-inflammatory cytokines as well as angiogenic factors to assess tumor aggressiveness.	Glioblastoma	IL1β IL2 IL6 IL8 IL10 TNFα VEGF FGF2 GMCSF	N/A	3-fold upregulation of IL6, IL1β, TNFα, IL10 2-fold upregulation of VEGF, FGF2, IL8, IL2 and GMCSF Cytokines are related to tumor progression,	The xMAP assay was utilized with Luminex 200 system to visualize plates. Cytokine and angiogenic factors were determined with the

Author (Reference)	Basics	Cancer Type	Gene or Gene Product	Gene Variability	Result	Procedure Used
					aggressiveness and disease-associated pain.	Human cytokine 12-plex. ELISA used to quantitate serum levels of growth factors.
Viet et al. 2011 (22)	Oral cancer patients – squamous cell carcinoma (n=20 patients) Compared to 8 oral dysplasia patients	Squamous cell carcinoma (pain) vs oral dysplasia	EDNRB	DNA methylation state	Oral SCC demonstrated higher methylation than oral dysplasia tissues at all queried CpG sites, were on average 20% higher; sig difference b/w methylation values at almost all CpG sites	Tissue samples
Oliviera et al. 2014 (23)	Analyze the relationship between TNF-alpha and C-reactive protein (CRP) and	Head and neck cancer	CRP TNFα	N/A	Positive associations were found between pain, CRP level and tumor stage.	Pain was determined using the Brief Pain Inventory Questionnaire.

Table 1. (Continued)

Author (Reference)	Basics	Cancer Type	Gene or Gene Product	Gene Variability	Result	Procedure Used
	perceived pain in head and neck cancer patients prior to cancer treatment.				Patients experiencing pain had high CRP and TNFα levels rather than patients experiencing no pain or controls.	Serum levels of CRP was evaluated using the particle-enhanced tabidimetric immunoassay (PETIA) and TNFα serum levels were quantitated using ELISA.
Berruti et al. 1999 (28)	n=323 patients with bone metastases; adenocarcinoma – off any systemic treatment for >1 month and palliative radiotherapy for >3 months	Breast, lung, prostate cancer	C-telopeptide of type I collagen (ICTP), serum carboxy-terminal deoxypyridinoline (DPYD), urinary molar ratio of calcium to creatinine (Ca:Cr), serum calcium (CaS),	N/A	Higher levels of ICTP= greater pain score ($p<0.001$) Higher levels of DPYD = greater pain score ($p=0.002$) Higher levels of CaS = greater pain score ($p=0.002$)	Urine and blood samples

Author (Reference)	Basics	Cancer Type	Gene or Gene Product	Gene Variability	Result	Procedure Used
			serum parathyroid hormone (PTH), bone-specific alkaline phosphatase (BALP),		Higher levels of Ca:Cr = greater pain score ($p=0.01$) Lower levels of PTH = lower pain score ($p=0.02$) Lung cancer, no measure DPYD, others all still significant Prostate cancer patients: BALP parallel bone pain ($p=0.05$), ICTP increased w pain score ($p<0.005$) Breast cancer patients: ICTP increased w pain score ($p<0.005$), Ca:Cr increased w increase of bone pain ($p=0.02$)	

Table 1. (Continued)

Author (Reference)	Basics	Cancer Type	Gene or Gene Product	Gene Variability	Result	Procedure Used
Scarpi et al. 2014 (29)	Examined differences in patients experiencing cancer induced bone pain (CIBP) verses those who do not. Examined differences in opioid response.	Tumors sites include: urologic, lung, breast, prostate, GI, female reproductive organs, hematological, unknown origin, other metastatic sites include: liver, bone, CNS, lung, other	*HTR3E* *HTR3B* (did not pass Benjamini-Hochberg criterion)	*HTR3E*: rs7432211, rs4912522 *HTR3B*: rs7103572, rs1176744	Haplotypes with an initial significant association with pain relief response reported here. None of the examined haplotypes were significantly associated with CIBP versus non-CIBP patients, or with pain relief response versus non-response.	Whole blood samples, DNA extraction and SNP analyses; haplotype analysis software
Oliviera et al. 2014 (30)	Analyzed the impact of *IL1A, IL1B* and *IL1RN* polymorphisms on cytokine serum levels and association with pain intensity.	Tumor sites include: lung, urologic, breast, prostate, multiple myeloma, female reproductive organs, and other	*IL1A* *IL1B* *IL1RA*	*IL1A*: rs17561 *IL1B*: rs1143634 *IL1RA*: rs419598	Homozygotes for the *T*-allele in the *IL1B* rs1143634 polymorphism were correlated with lower IL1β levels and lower pain levels. Female	Serum compounds assessed using ELISA. Patients genotyped using real-time PCR. Brief

Author (Reference)	Basics	Cancer Type	Gene or Gene Product	Gene Variability	Result	Procedure Used
	Performance status, adverse events, breakthrough pain and metastasis in 74 Caucasian cancer patients	Metastatic sites include: None, liver, bone, CNS, lung, non-visceral (unknown location), visceral (unknown location)			*T* homozygotes with metastatic disease had lower IL1β levels. High IL1β levels were linked to a 4-fold increased probability of metastatic disease.	Pain Inventory Questionnaire was used to evaluate pain scores. Results were compared to pain parameters as well as CNS images.
Heitzer et al. 2012 (31)	Examined serum cytokines correlating with pain intensity in patients with tumor-caused advanced cancer pain.	Various (n=19)	IL7 IL18 MCP1 MIPIβ OPG	N/A	Decreases in serum concentrations of IL7, IL18, MCP1, MIPIβ and OPG were correlated with targeted pain relief with morphine. OPG serum levels were higher in patients with nociceptive pain versus those with visceral or bone pain.	Serum cytokine concentrations were assessed before and during morphine treatment and then assessed relationships between pain entities, intensity, gender, primary tumor location, and

Table 1. (Continued)

Author (Reference)	Basics	Cancer Type	Gene or Gene Product	Gene Variability	Result	Procedure Used
						baseline cytokine concentrations. Pain entities were evaluated via the Brief Pain Inventory and the Short Form McGill Pain Questionnaire. Cytokines analyzed via ELISA.
Rokyta et al. 2009 (32)	Analyzing biochemical markers that were descriptive for pain processes	Various	Serum albumin HDL cholesterol Apolipoprotein A CRP Alpha1 globulin Glucose	N/A	Pre-therapy oncological patients had lower levels of serum albumin, HDL cholesterol and apolipoprotein A, while CRP and alpha1 globulin	Blood samples, pain questionnaires. Oncological patients received NSA, medium strength and strong opioids,

Author (Reference)	Basics	Cancer Type	Gene or Gene Product	Gene Variability	Result	Procedure Used
	n=40 patients (20 oncological patients and 20 non-oncological patients)				were higher compared to non-oncological patients. After therapy, glucose and alpha1 globulin were significantly higher and levels of apolipoprotein A were lower in oncological patients than non-oncological patients. Irrespective of diagnosis, patients given antidepressants had higher levels of gamma globulin compared to non-treated.	and antidepressants. Non-oncological patients were given NSA and weak opioids.

Another study by Langford et al. utilized the same patient population and analyzed the impact of inter-individual differences in potassium channels on pain scores prior to surgery. Out of 155 SNPs in 10 candidate potassium channel genes, variations in *KCNS1, KCNJ3, KCNJ6*, and *KCNK9* were found to be linked to breast cancer pain (10).

Stephens et al. conducted a longitudinal study analyzing the influence of genetic variation of cytokine genes on breast cancer pain. A sample of 398 women completed the Breast Symptoms Questionnaire and Postsurgical Pain Questionnaire to assess preparative and acute postoperative pain respectively. The patient cohort was further subdivided into no pain and pain groups. Following adjustment for preoperative breast pain and the severity of average postoperative breast pain, carriers of the *IL1R2* SNP rs11674595 and the *IL10* haplotype A8 (composed of seven SNPs: rs3024505C, rs3024498G, rs3024496C, rs1878672G, rs1518111A, rs1518110T, rs3024491T) were more likely to belong to the pain group. Indeed, patients homozygous for the "C" allele in *IL1R2* rs11674595 had a 36.1-fold increase of experiencing severe pain following surgery (p=0.015). Each dose of *IL10* haplotype A8 decreased the chance of being a member of the severe breast cancer pain group by 79% (p=0.037) (11).

Lung cancer

Reyes-Gibby et al. (2007) explored the association between pain severity in 606 non-small cell lung cancer (NSCLC) patients of all stages (446 White, 125 African-American, and 35 Hispanic patients) and genetic variation in candidate cytokine genes including *tumour necrosis factor-α* (*TNF 308 G/A*), *interleukin 6* (*IL6*) *-174G/C,* and *IL8 -251T/A.* The A allele in *IL8* was associated with increased pain severity in white patients (p=0.04). The significance of this relationship persisted after controlling for epidemiologic, clinical and symptom variables known to influence pain phenotypes (OR=2.35, p=0.03). However, no significant association between *TNF*, *IL6*, and *IL8* genotypes and severe pain was observed among African-Americans and Hispanics (12).

Reyes-Gibby et al. (13) examined the influence of cytokine gene variants in predicting pain severity in relation to morphine uptake in 140 lung cancer patients. *TNF -308G/A* and *IL6 -174G/C* were significantly associated with both pain severity and morphine equivalent daily dose. Carriers of the *IL6 -174CC* genotype needed to consume 4.7x more opioids to achieve analgesia compared to homozygotes and heterozygotes of the G allele (13).

Reyes Gibby et al. (2009) analyzed 59 SNPs in 37 inflammatory genes in 667 non-Hispanic Caucasian patients. Carriers of the CC genotype in *PTGS2* exon *10+837T>C* (rs5275) were found to have a reduced risk of severe pain (OR=0.33). An additive model for *TNF -308GA* (rs1800629) (OR=1.67) and *NFKBIA Ex6+50C>T* (rs8904) was predictive of severe pain (OR=0.64). In a multi-gene analysis, a gene-dose effect was found, with each protective genotype reducing the risk for severe pain by as much as 38% (14).

Reyes-Gibby et al. (15) assessed the relationship of 55 SNPs in immune response genes and the occurrence of severe pain in 599 Caucasians with untreated and histologically confirmed NSCLC. In this patient sample, 116 patients were in the severe symptom cluster (high pain intensity, depression and fatigue), while 183 patients were in the low symptom cluster. An additive combination of 5 SNPs that comprised of *ENOS -1474 T/A* (OR=0.54), *IL1B T-31C* (OR = 0.55), *TNFR2 Met196Arg* (OR=1.85), *PTGS2* exon *10+837T>C* (OR= 0.54), and *IL10RB Lys47Glu* (OR=1.74) predicted high symptom burden (15). In another analysis of this study population, Reyes-Gibby et al. (16) reported the *IL7 -T251A* SNP to be the most relevant genetic factor for pain (OR=2.18), depressed mood (OR=0.37) and fatigue (OR=2.07). Female patients with *Lys_Glu* or *Glu_Glu* in the *IL10RB* Lys^{47}*Glu* variation had a 0.49x lower risk of severe fatigue compared to carriers of Lys_Lys (P=0.027). Male subjects with *IL1A C-889T*, *CT* or *TT* genotypes had lower risk of fatigue compared with *CC* genotype (OR=0.38). This study proposes that targeting the *IL10* and *IL8* genes are putative targets for treating multiple cancer symptoms (16).

Rausch et al. examined the effects of 10 SNPs from the *LTA* and *PTGS2* inflammatory genes on cancer pain in lung cancer survivors. A total of 1,149 Caucasian lung cancer patients were genotyped and completed the Short-Form General Health Survey (SF-8) to determine QOL parameters. This sample was further subdivided into three groups based on survivorship: early survivors (<3 years), middle term survivors (3-5 years), and long term survivors (>5 years following lung cancer diagnosis). Utilizing multiple regression modeling, significant relationships were found between 3 SNPs and SF-8 parameters. Homozygous or heterozygous early survivor carriers of the minor G allele in *PTGS2* rs5277 were associated with high pain scores. Patients with one or more of the G allele in *LTA* rs1799964 were associated with reduced pain scores (17).

Wang et al. evaluated biological markers associated with developing symptom burden in 62 patients undergoing chemoradiation therapy (18). An increase of IL6 serum levels was associated with an increase in severity of 5 symptoms (pain, fatigue, disturbed sleep, lack of appetite and sore throat) (18).

An increase from baseline in IL6 concentration was associated with a significant increase in mean severity of 5 severe symptoms (18). During chemoradiation therapy, increased serum TNF-R1 was the only cytokine related with a higher 15 symptom component score (18).

Prostate cancer

Tagawa et al. utilized circulating tumour cell counts (CTC) from 62 patients as a real-time tool to analyze the sensitivity of taxanes in chemo-naïve metastatic castration-resistant prostate cancer. Taxanes are the only chemotherapeutic with a survival benefit in prostate cancer and the employment of multiplex confocal microscopy to observe androgen receptor localization potentially related to taxane efficacy or resistance. The study noted higher CTC levels indicative of higher baseline LDH, pain score (p=0.0031) and poor prognostic factors (19).

Pancreatic cancer

Reyes-Gibby et al. (20) explored the association between SNPs in cytokine genes (*IL1B, IL6, IL8, IL10, IL18, TNF, NFKB1*) and pain severity for 156 Caucasian pancreatic cancer patients. Carriers of the AT or TT genotypes in *IL8-251T/A* were associated with a 3-fold risk of severe pain relative to AA genotypes (OR = 3.23, 95% CI = 1.4, 4.7) (20).

Hepatobiliary carcinoma

Steel et al. examined the relationship between alterations in immunity and symptoms in 206 cancer patients (hepatocellular carcinoma (84%), gallbladder carcinoma (6%), cholangiocarcinoma (5%), or neuroendocrine carcinoma (3%), and other primary tumours with liver metastases (2%)). Patients completed the Functional Assessment of Cancer Therapy-Hepatobiliary (FACT-Hep) questionnaire to assess symptom change and treatment side effects. Peripheral blood leukocytes were measured at baseline and at 3 and 6-month intervals. Patients were further subdivided into symptomatic, asymptomatic and fatigue clusters. Significantly higher eosinophils were found in the symptomatic cluster at 3 (3.1 vs. 2.9 and 1.6) and 6-month (3.6 vs. 3.0 and 1.5) intervals in comparison to the asymptomatic and fatigued clusters, respectively. A trend toward significance for eosinophils and FACT-Pain change (p=0.056) was observed. Correlation was also found between immune system parameters and the rate of change in symptomology. Only hepatocellular carcinoma produced the same result in post hoc analysis with eosinophils and pain being significantly related over time (21).

Head and neck cancer

Viet et al. compared the methylation state of the *EDNRB* gene in 20 oral squamous cell carcinoma (SCC) and 8 oral dysplasia patients. *EDNRB* encodes the endothelin B (ET_B) receptor and is often hypermethylated and therefore silenced in various neoplastic lesions (22). Although the role of ET_B in nociception is uncertain, *EDNRB* is heavily methylated in painful human oral SCC lesions and not methylated in oral dysplasia. On average, oral SCC biopsies had 20% higher methylation at all queried CpG sites than oral dysplasia tissues. This resulted in 19 of 20 oral SCC samples to have a 1.15 to 53.0-fold change in *EDNRB* mRNA levels compared to normal tissues from the same patients (22).

Oliveira et al. indicated that head and neck cancer patients with higher serum levels of C-reactive protein (CRP) and TNFα were more likely to experience pain compared controls and patients not expressing painful symptoms (23).

Glioblastoma

Albulescu et al. profiled 55 confirmed glioblastoma patients for serum pro- and anti-inflammatory cytokines and angiogenic factors to assess tumour aggressiveness and pain. Results showed that dysregulation of cytokines and angiogenic factors was associated with pain. Glioblastoma patients had an over 3-fold up-regulation of IL6, IL1B, TNFα, and IL10, and 2-fold up-regulation of VEGF, FGF2, IL8, IL2, and GMCSF compared to healthy controls. The expression pattern of these cytokines was correlated with tumour grade, proliferative markers, clinical aggressiveness, and generation of disease-associated pain (24).

Myeloproliferative neoplasms

Nguyen et al. (25) performed genomic analysis of the *JAK2 V617F* mutation to quantitatively compare the symptom burden of patients with chronic myelogenous leukemia (CML) and systemic mastocytosis (SM) with other forms of myeloproliferative neoplasms (MPNs), including polycythemia vera (PV), essential thrombocythemia (ET), and primary myelofibrosis (PMF). Results from a patient-completed MPN-symptom assessment form found that the *V617F* mutation was significantly correlated to bone pain (p=0.0341), abdominal pain and discomfort (p=0.0371), and pruritus (p=0.0241) (25).

A study by Liu et al. (26) sought to examine whether SNPs in the *TLC1A* gene alter cytokine expression, and also understand potential mechanisms behind aromatase inhibitor-induced side effects in breast cancer patients. 300 lymphoid blastoid lines from 100 from European-American, 100 African-American, and 100 Chinese-American blood samples were used in this study. Experiments in vitro involving addition of an estradiol antagonist (mimicking aromatase inhibitors) in the presence of increased estradiol led to increased activity levels of NFκB in variant lines, which may be responsible for the aromatase-induced side effects such as arthralgias and myalgias (26).

Advanced osteolytic bone sisease

Engler et al. (27) examined the effect of pamidronate therapy on bone resorption markers in 70 cancer patients with malignant osteolytic bone disease and how these resorption parameters were linked to clinical outcomes and symptoms. Lower levels of serum C-telopeptide of type I collagen (ICTP) and urinary free deoxypyridinoline (DYD) significantly predicted decreased pain intensity and Serum alkaline phosphatase was also predictive of this parameter (27).

Other neoplastic sites

Berruti et al. (28) searched for bone marker patterns associated with primary tumour type, bone pain and number of cancer sites in 323 patients with bone metastases from various primary malignancies. Higher levels of ICTP, DPYD, CaS and the Ca:Cr ratio were significantly associated with higher pain scores. Patients with lower levels of parathyroid hormones had lower pain scores (p=0.02). All measures, except for DPYD, were significant for lung cancer patients. Serum ICTP levels increased with pain in both breast and prostate cancer patients (p<0.005), while bone alkaline phosphatase (BALP) levels paralleled the increase in bone pain in prostate cancer patients (p=0.05). Similarly, increasing Ca:Cr ratio was associated with increased bone pain in breast cancer patients (p=0.02) (28).

A study by Scarpi et al. (29) looked for genetic and clinical differences between Caucasian patients experiencing cancer-induced bone pain (CIBP) and non-CIBP, as well as good versus poor opioid response in CIBP patients. CIBP patients had more breakthrough cancer pain episodes, more pain interference with walking ability, used more adjuvant drugs, and had higher (albeit non-significant) overall survival than non-CIBP. However, no specific genetic patterns emerged between CIBP vs non-CIBP or responsive vs nonresponsive CIBP patients (29).

Oliveira et al. (30) investigated the association of SNPs in *IL1A, IL1B and IL1RN* on serum cytokine levels, pain intensity, performance status, adverse effects, metastases, and breakthrough pain in 74 Caucasian cancer patients. Carriers homozygous for the T-allele of *ILIB* rs1143634 were associated with reduced IL1B levels and lower pain scores. Elevated IL1B levels were in turn linked with a 4-fold increase in the likelihood of metastatic disease (30).

Heitzer et al. (31) evaluated 19 biomarkers in 20 healthy control patients who experienced no pain and 20 cancer patients with pain. Participants completed a numeric rating scale (NRS) prior to and following morphine treatment in order to determine changes in pain scores. Five biomarkers, including IL7, IL8, MCP1, MIP1α, MIP1β, and OPG were significantly correlated with pain relief (31).

Rokyta et al. (32) examined the biochemical markers associated with pain processes in 20 cancer patients and 20 non-cancer patients. Oncological patients had lower levels of serum albumin, HDL cholesterol and apolipoprotein A, and higher levels of CRP and alpha1 globulin in comparison to non-oncological patients. After therapy, glucose and alpha1 globulin were significantly higher and levels of apolipoprotein A were lower in oncological patients than non-oncological patients. Irrespective of diagnosis, patients given antidepressants had higher levels of gamma globulin compared to patients without antidepressants (32).

DISCUSSION

The study of genetic variations and biological markers associated with cancer-related pain sensitivity holds the potential of informing dosing protocols and upgrading best practices in oncology with the ultimate goal of improving healthcare outcomes. The increasing number of cancer diagnoses and survivors emphasizes the importance of reducing the pain prevalence at all stages of the disease (33).

This chapter is representative of the expansive body of literature linking cytokine levels and SNPs with cancer symptoms. The elucidation and understanding of the underlying causes of symptom evolution and burden would be beneficial for developing protocols and practices to better control adverse treatment and disease-related effects. Identifying and studying the biological mechanisms like the inflammatory process that may account for the formation and continuation of debilitating cancer symptoms like pain could inform the production of initiatives and innovations to ameliorate those processes.

Although there are numerous mechanisms of cancer pain, inflammation likely plays a central role. Elevated levels of circulating inflammatory and immunological markers have been linked to shorter survival, event-free survival and remission (36). This review of the literature also demonstrates that inflammation markers are also predictors of symptom burden and QOL in multiple forms of cancer. A review of the literature points to the role of pro-inflammatory cytokines like interleukins IL1, IL2, IL6, IL8, TNFα, and interferon γ (IFNγ) in nociceptive transmission, analgesia and neuropathic pain (24, 34, 35).

The molecular mechanism underlying the impact of cytokines on pain response is not completely understood. A disruption of tissue homeostasis and repair common to both inflammation and cancer leads to cytokine release, altering nociceptor activity (37). Animal models have demonstrated IL8 causes dose dependent hyperalsia while other reports have highlighted the contribution of IL6 and TNFα to hyperexcitability in neural pain transmission (12, 38, 39). These correlations suggest a potential polygenic model of cytokines and biological processes in the modulation of nociception in cancer.

In addition to societal opposition to opioid use and the genetic variables and biological markers presented herein, inter-individual differences in genes pertaining to the pharmacokinetics and pharmacodynamics of opioids impact pain-sensitivity. A meta-analysis conducted by Ren et al. reported the *A118G* allele variant of *OPRM1* has one of the greatest influences on pain management of postoperative cancer pain. Carriers of this SNP required more opioids to attain an analgesic response ($p<0.00001$) and reported higher pain scores in the first 24-hours of the postoperative period. Compared to homozygotic *CYP3 A4*1/*1* patients, Carriers of the *CYP3 A4*1G* SNP were associated with reduced opioid consumption in the first 24-hours following surgery. Genetic information pertaining to the effect of analgesic treatment can augment cancer pain management (40).

The WHO published a three-step analgesic ladder (1986) to advise practitioners on the management of cancer-related pain, which has successfully relieved pain in 75% of patients (41). This recommendation included the treatment of global cancer pain using oral analgesics in a clock-regulated manner (1). Despite the high prevalence of cancer-related pain and the availability of pharmacologic therapies, pain is often inadequately managed in cancer patients, primarily in children, seniors, metastatic outpatients and minority groups (42-45). Misconceptions regarding opioid provision is the most cited reason of under-treatment and pain mismanagement. A retrospective study conducted by Jain et al. found that of the 31.5% of cancer patients referred to the Tata Memorial Hospital pain clinic with severe pain, only 14.4% of this sample were treated with opioids, while 69% never received any analgesics at diagnosis. Approximately 63.7% achieved analgesia following the initiation of opioid and adjuvant treatment (1). A survey of 4,618 physicians in British Columbia by Gallagher et al. stated that oncologists had the highest and surgeons had the lowest knowledge of cancer pain management. The most frequent incorrectly answered questions regarded equianalgesic (68%) and adequate breakthrough dosing (45%), highlighting gaps between what is known about pain management and actual practices of healthcare professionals (46). Although opioids are the pillar of moderate to severe cancer pain management, phobias of prescribing opioids, prescriptions of opioid only on an 'as needed basis, patient-specific barriers to opioid treatment (such as opioid addiction), a misunderstanding of adequate pain treatment, lack of compliance, stigmatization, and an inability to recognize specific pain syndromes all contribute to the misuse of opioids (1).

The National Cancer Control Network (NCCN) advocates for a multi-dimensional approach to cancer therapy where each patient is evaluated individually based on disease stage, treatment response, and provision of therapeutics in a flexible manner (1). Addressing the biopsychosocial needs of the patient is also a requirement to successfully treat pain. Advancements in early diagnosis and treatment of cancer have led to better prognostic outcomes, including a higher likelihood of survival (48). Following treatment and remission, somatic and psychological burdens often negate the possibility of a normal life and autonomy. For this reason, survivors require support systems to cope with impairments caused by disease (49). A study conducted by Lamprecht et al. examined the benefits of oncological rehabilitation on factors associated with health-related QOL in breast, prostate and colon cancer patients (50). Colon cancer patients improved the most in global QOL while prostate cancer patients improved in emotional functioning. Symptom and fatigue burden also improved following rehabilitation. Although the functional burdens improved in the 3-month period following rehabilitation, the emotional functioning of breast cancer patients declined significantly (50). This result substantiates the requirement of appropriate emotional and anxiety support for cancer patients following discharge. Additionally, the risk of unemployment for cancer patients is particularly high when compared to the healthy population (51), as symptom burden and side effects of treatment, including pain, can exacerbate a person's performance in the workplace. Initiatives such as the Cancer Survivorship program provides patients assistance for the first 5 years following therapy and the American Cancer Society's "reach to recovery" provides volunteers that act as role models and support systems for cancer patients and can aid in the achievement of this goal (50).

In summation, the study of genetic and biological markers outlined in this chapter defines the need for statistically powerful gene-wide associated studies (GWAS) with the purpose of establishing a personalized form of treatment. The literature, as it applies to cancer-related pain sensitivity, illuminates novel targets for combatting one of the most prevalent and poignant symptoms of cancer, and augments the desire for more conclusive research for the benefit of health care providers, patients and the field of oncology as a whole.

ACKNOWLEDGMENT

We thank the generous support of Joey and Mary Furfari Cancer Research Fund.

REFERENCES

[1] Jain PN, Pai K, Chatterjee AS. The prevalence of severe pain, its etiopathological characteristics and treatment profile of patients referred to a tertiary cancer care pain clinic. *Indian J Palliat Care* 2015;21(2):148.

[2] Cherny NI, Coyle N, Foley KM. Suffering in the advanced cancer patient: a definition and taxonomy. *J Palliat Care* 1994;10(2):57-70.

[3] Breivik H, Cherny N, Collett B, De Conno F, Filbet M, Foubert AJ, et al. Cancer-related pain: a pan-European survey of prevalence, treatment, and patient attitudes. *Ann Oncol* 2009;20(8):1420-33.

[4] Maier SF, Watkins LR. Immune-to-central nervous system communication and its role in modulating pain and cognition: Implications for cancer and cancer treatment. *Brain Behav Immun* 2003;17(1):125-31.

[5] Thorne C. Management of arthralgias associated with aromatase inhibitor therapy. *Curr Oncol* 2007;14(Suppl 1):S11-9.

[6] McDonald R, Chow E, Rowbottom L, DeAngelis C, Soliman H. Incidence of pain flare in radiation treatment of bone metastases: A literature review. *J Bone Oncol* 2014;3(3-4):84-9.

[7] Young EE, Lariviere WR, Belfer I. Genetic Basis of Pain Variability: Recent Advances. *J Med Genet* 2012;49(1):1-18.

[8] Lengacher C, Reich R, Paterson C, Shelton M, Ramesar S, Alinat C, et al. The relationship between stress hormones (cortisol), cytokines (IL-6) and symptoms among breast cancer survivors (BCS) in an MBSR (BC) randomized controlled trial (RCT). *Psychooncol* 2015;24:95-6.

[9] McCann B, Miaskowski C, Koetters T, Baggott C, West C, Levine JD, et al. Associations between pro-and anti-inflammatory cytokine genes and breast pain in women prior to breast cancer surgery. *J Pain* 2012;13(5):425-37.

[10] Langford DJ, West C, Elboim C, Cooper BA, Abrams G, Paul SM, et al. Variations in potassium channel genes are associated with breast pain in women prior to breast cancer surgery. *J Neurogenet* 2014;28(1-2):122-35.

[11] Stephens K, Cooper BA, West C, Paul SM, Baggott CR, Merriman JD, et al. Associations between cytokine gene variations and severe persistent breast pain in women following breast cancer surgery. *J Pain* 2014;15(2):169-80.

[12] Reyes-Gibby CC, Spitz M, Wu X, Merriman K, Etzel C, Bruera E, et al. Cytokine genes and pain severity in lung cancer: exploring the influence of TNF-α-308 G/A IL6-174G/C and IL8-251T/A. *Cancer Epidemiol Biomarkers Prev* 2007;16 (12):2745-51.

[13] Reyes-Gibby CC, El Osta B, Spitz MR, Parsons H, Kurzrock R, Wu X, et al. The influence of tumor necrosis factor-α− 308 G/A and IL-6− 174 G/C on pain and analgesia response in lung cancer patients receiving supportive care. *Cancer Epidemiol Biomarkers Prev* 2008;17(11):3262-7.

[14] Reyes-Gibby CC, Spitz MR, Yennurajalingam S, Swartz M, Gu J, Wu X, et al. Role of inflammation gene polymorphisms on pain severity in lung cancer patients. *Cancer Epidemiol Biomarkers Prev* 2009;18(10):2636-42.

[15] Reyes-Gibby CC, Swartz MD, Yu X, Wu X, Yennurajalingam S, Anderson KO, et al. Symptom clusters of pain, depressed mood, and fatigue in lung cancer: assessing the role of cytokine genes. *Support Care Cancer* 2013;21(11):3117-25.

[16] Reyes-Gibby CC, Wang J, Spitz M, Wu X, Yennurajalingam S, Shete S. Genetic variations in interleukin-8 and interleukin-10 are associated with pain, depressed mood, and fatigue in lung cancer patients. *J Pain Symptom* 2013;46(2):161-72.

[17] Rausch SM, Gonzalez BD, Clark MM, Patten C, Felten S, Liu H, et al. SNPs in PTGS2 and LTA predict pain and quality of life in long term lung cancer survivors. *Lung Cancer* 2012;77(1):217-23.

[18] Wang XS, Shi Q, Williams LA, Mao L, Cleeland CS, Komaki RR, et al. Inflammatory cytokines are associated with the development of symptom burden in patients with NSCLC undergoing concurrent chemoradiation therapy. *Brain Behav Immun* 2010;24(6):968-74.

[19] Tagawa ST, Galletti G, Antonarakis ES, Tasaki S, Gjyrezi A, Worroll D, et al. Baseline analysis of circulating tumor cell (CTC) enumeration and androgen receptor (AR) localization in men with metastatic castration-resistant prostate cancer (mCRPC) in taxynergy. *ASCO* 2015;33(15 Suppl):5031.

[20] Reyes-Gibby CC, Shete S, Yennurajalingam S, Frazier M, Bruera E, Kurzrock R, et al. Genetic and nongenetic covariates of pain severity in patients with adenocarcinoma of the pancreas: assessing the influence of cytokine genes. *J Pain Symptom Manage* 2009;38(6):894-902.

[21] Steel JL, Kim KH, Dew MA, Unruh ML, Antoni MH, Olek MC, et al. Cancer-related symptom clusters, eosinophils, and survival in hepatobiliary cancer: an exploratory study. J Pain Symptom Manage 2010;39(5):859-71.

[22] Viet CT, Ye Y, Dang D, Lam DK, Achdjian S, Zhang J, et al. Re-expression of the methylated EDNRB gene in oral squamous cell carcinoma attenuates cancer-induced pain. *Pain* 2011;152(10):2323-32.

[23] Oliveira KG, von Zeidler SV, Lamas AZ, de Podestá JR, Sena A, Souza ED, et al. Relationship of inflammatory markers and pain in patients with head and neck cancer prior to anticancer therapy. Braz J Med Biol Res 2014;47(7):600-4.

[24] Albulescu R, Codrici E, Popescu ID, Mihai S, Necula LG, Petrescu D, et al. Cytokine patterns in brain tumour progression. *Mediators Inflamm* 2013;2013:979748.

[25] Nguyen HM, Hinds DA, Barnholt KE, Kiefer AK, Do CB, Eriksson N, et al. Correlation of symptom assessment with genotyping analysis of saliva samples in a large cohort of myeloproliferative neoplasm patients. *Blood* 2012;120(21):1732.

[26] Liu M, Wang L, Bongartz T, Hawse JR, Markovic SN, Schaid DJ, et al. Aromatase inhibitors, estrogens and musculoskeletal pain: estrogen-dependent T-cell leukemia 1A (TCL1A) gene-mediated regulation of cytokine expression. Breast Cancer Res 2012;14(2):R41.

[27] Engler H, Koeberle D, Thuerlimann B, Senn HJ, Riesen WF. Diagnostic and prognostic value of biochemical markers in malignant bone disease: a prospective study on the effect of bisphosphonate on pain intensity and progression of malignant bone disease. *Clin Chem Lab Med* 1998;36(11):879-85.

[28] Berruti A, Dogliotti L, Gorzegno G, Torta M, Tampellini M, Tucci M, et al. Differential patterns of bone turnover in relation to bone pain and disease extent in bone in cancer patients with skeletal metastases. *Clin Chem* 1999;45(8):1240-7.

[29] Scarpi E, Calistri D, Klepstad P, Kaasa S, Skorpen F, Habberstad R, et al. Clinical and genetic factors related to cancer-induced bone pain and bone pain relief. *Oncologist* 2014;19(12):1276-83.

[30] Oliveira A, Dinis-Oliveira RJ, Nogueira A, Gonçalves F, Silva P, Vieira C, et al. Interleukin-1β genotype and circulating levels in cancer patients: metastatic status and pain perception. *Clin Biochem* 2014;47(13-14):1209-13.

[31] Heitzer E, Sandner-Kiesling A, Schippinger W, Stohscheer I, Osprian I, Bitsche S, et al. IL-7, IL-18, MCP-1, MIP1-β, and OPG as biomarkers for pain treatment response in patients with cancer. *Pain Physician* 2012;15(6):499-510.

[32] Rokyta R, Haklova O, Yamamotova A. Assessment of chronic benign and cancer pain using blood plasma biomarkers. Neuro Endocrinol Lett 2009;30(5):637-42.

[33] Van den Beuken-van Everdingen MH, De Rijke JM, Kessels AG, Schouten HC, Van Kleef M, Patijn J. Prevalence of pain in patients with cancer: a systematic review of the past 40 years. *Ann Oncol* 2007;18(9):1437-49.

[34] Hutchinson MR, Coats BD, Lewis SS, Zhang Y, Sprunger DB, Rezvani N, et al. Proinflammatory cytokines oppose opioid-induced acute and chronic analgesia. *Brain Behav Immun* 2008;22(8):1178-89.

[35] Kawasaki Y, Xu ZZ, Wang X, Park JY, Zhuang ZY, Tan PH, et al. Distinct roles of matrix metalloproteases in the early-and late-phase development of neuropathic pain. *Nat Med* 2008;14(3):331-6.

[36] Seruga B, Zhang H, Bernstein LJ, Tannock IF. Cytokines and their relationship to the symptoms and outcome of cancer. Nat Rev Cancer 2008;8(11):887-99.

[37] Brisby H, Olmarker K, Larsson K, Nutu M, Rydevik B. Proinflammatory cytokines in cerebrospinal fluid and serum in patients with disc herniation and sciatica. *Eur Spine J* 2002;11(1):62-6.

[38] Cunha FQ, Lorenzetti BB, Poole S, Ferreira SH. Interleukin□8 as a mediator of sympathetic pain. *Br J Pharmacol* 1991;104(3):765-7.

[39] Maier SF, Watkins LR. Cytokines for psychologists: implications of bidirectional immune-to-brain communication for understanding behavior, mood, and cognition. *Psychol Rev.* 1998;105(1):83-107.

[40] Lu L. The impact of genetic variation on sensitivity to opioid analgesics in patients with postoperative pain: a systematic review and meta-analysis. *Pain physician* 2015;18(2):131-52.

[41] Zech DF, Grond S, Lynch J, Hertel D, Lehmann KA. Validation of World Health Organization Guidelines for cancer pain relief: a 10-year prospective study. *Pain* 1995;63(1):65-76.

[42] Bernabei R, Gambassi G, Lapane K, Landi F, Gatsonis C, Dunlop R, et al. Management of pain in elderly patients with cancer. SAGE Study Group. Systematic Assessment of Geriatric Drug Use via Epidemiology. *JAMA* 1998;279(23):1877-82.

[43] Cleeland CS, Gonin R, Hatfield AK, Edmonson JH, Blum RH, Stewart JA, et al. Pain and its treatment in outpatients with metastatic cancer. *N Engl J Med* 1994;330(9):592-6.

[44] Wolfe J, Grier HE, Klar N, Levin SB, Ellenbogen JM, Salem-Schatz S, Emanuel EJ, Weeks JC. Symptoms and suffering at the end of life in children with cancer. *N Engl J Med* 2000;342(5):326-33.

[45] Cleeland CS, Gonin R, Baez L, Loehrer P, Pandya KJ. Pain and treatment of pain in minority patients with cancer: the Eastern Cooperative Oncology Group Minority Outpatient Pain Study. *Ann Intern Med* 1997;127(9):813-6.

[46] Gallagher R, Hawley P, Yeomans W. A survey of cancer pain management knowledge and attitudes of British Columbian physicians. *Pain Res Manag* 2004;9(4):188-94.

[47] Cohen MZ, Easley MK, Ellis C, Hughes B, Ownby K, Rashad BG, et al. Cancer pain management and the JCAHO's pain standards: an institutional challenge. *J Pain Symptom Manage* 2003;25(6):519-27.

[48] Aaronson NK, Mattioli V, Minton O, Weis J, Johansen C, Dalton SO, et al. Beyond treatment–psychosocial and behavioural issues in cancer survivorship research and practice. *EJC Suppl* 2014;12(1):54-64.

[49] Handberg C, Nielsen CV, Lomborg K. Men's reflections on participating in cancer rehabilitation: a systematic review of qualitative studies 2000–2013. *Eur J Cancer Care* (Engl) 2014;23(2):159-72.

[50] Lamprecht J, Thyrolf A, Mau W. Health□related quality of life in rehabilitants with different cancer entities. *Eur J Cancer Care* (Engl) 2016; DOI: 10.1111/ecc.12554.

[51] De Boer AG, Taskila T, Ojajärvi A, van Dijk FJ, Verbeek JH. Cancer survivors and unemployment: a meta-analysis and meta-regression. *JAMA* 2009;301(7):753-62.

In: Oncology
Editors: A. Furfari et al.

ISBN: 978-1-53610-608-4

Chapter 3

BIOMARKERS, CANCER-RELATED SYMPTOM BURDEN AND QUALITY OF LIFE

Leigha Rowbottom[1], MD(C), Stephanie Chan[1], BSc(C), Bo Angela Wan[1], MPhil, Pearl Zaki[1], BSc(C), Matthew Choi[1], BSc, Anthony Furfari[1], MD(C), Rachel McDonald[1], MD(C), Carlo DeAngelis[1], PharmD, Azar Azad[2,3,4], PhD, Ronald Chow[1], BMSc(C), Henry Lam[1], MLS, Edward Chow[1], MBBS and George S Charames[3,4,5,*], PhD

[1]Odette Cancer Centre, Sunnybrook Health Sciences Centre, University of Toronto, Toronto, Ontario, Canada
[2]Mount Sinai Services Inc., Toronto, Ontario, Canada
[3]Pathology and Laboratory Medicine, Mount Sinai Hospital, Toronto, Ontario, Canada
[4]Laboratory Medicine and Pathobiology, University of Toronto, Toronto, Ontario, Canada
[5]Lunenfeld-Tanenbaum Research Institute, Sinai Health System, Toronto, Ontario, Canada

---

[*] Correspondence: George S Charames PhD, Pathology and Laboratory Medicine, Mount Sinai Hospital, 600 University Ave, Toronto, Ontario, Canada. E-mail: George.Charames@sinaihealthsystem.ca.

Cancer and associated treatments often have a significant impact on patients and can manifest in a variety of symptoms leading to negative outcomes on health-related quality of life (HRQOL). Quite commonly certain symptoms may occur in conjunction, also referred to as a symptom cluster, which can lead to aggregate burden on the patient. Symptom clusters may occur together as there may be one true underlying etiology, which if identified could provide therapeutic targets for enhanced symptom management. Often improvement in one symptom can improve patient experience with other symptoms eventually leading to reduced overall symptom burden and improved HRQOL. Biomarkers may provide clues to physicians as to what the cause of a symptom or symptom cluster is, as well as to help identify symptoms in non-communicative patients. The objective of this chapter was to summarize the available literature regarding biological and genetic biomarkers related to cancer-related symptoms and quality of life (QOL) in cancer patients. A total of 41 studies were included in this review that reported on commonly experienced symptoms in cancer patients including fatigue, depression, anxiety, as well as QOL measures.

INTRODUCTION

Cancer and its associated treatments can cause a variety of distressing symptoms that can have a significant impact on affected patients. Both the presence and intensity of symptoms experienced can lead to significant burden on the patient, leading to diminished health-related quality of life (HRQOL), which has been previously observed across a variety of patient populations (1). HRQOL encompasses numerous dimensions of an individual's wellbeing including health, social and emotional well-being, cognition, role functioning, sexual functioning, physical functioning, symptoms, and spirituality (2). With the development of precision medicine, such patient reported holistic concepts will be of great utility when providing customized care to address individual concerns and goals.

Quite often certain symptoms may commonly occur in conjunction, which is then referred to as a symptom cluster. Symptom clusters can have aggregate negative effects on the patient and represent a unique challenge to physicians, as symptoms may be inter-related or due to one common etiology. For instance, a patient may be experiencing significant cancer-related pain that then leads to exacerbation of depressive symptoms, where treatment of the depressive symptoms alone may not achieve therapeutic efficacy. In such instances, a holistic approach using multimodal treatment options may better address the patient's overall symptom experience. The clinical utility of biomarkers is evident for symptom clusters as they can be used to predict or monitor multiple distress symptoms as disease progresses or as a patient undergoes treatment (3). Previous studies investigating symptom clusters in chemotherapy patients have found that pre-treatment clusters are associated with increased symptoms during treatment. Interventions directed at these symptoms can be implemented to reduce the prevalence and intensity during treatment, which is a time when patients are particularly sensitive to the impact of various symptoms (4). Biomarkers could be potentially used to identify which patients are at risk of high symptom burden on treatment and help to educate patients on what they as individuals can expect and provide prophylactic interventions. Not only do biomarkers have the potential be used to predict patient symptoms, but may even be used as therapeutic targets.

In this chapter we summarize the literature regarding biomarkers associated with various cancer-related symptoms. Symptoms are experienced throughout the entire cancer trajectory and represent a large burden on the patient and healthcare system. In the future, biomarkers may assist healthcare professionals in the appropriate management of symptoms in order to improve outcomes for cancer patients, particularly those at the end of life.

OUR LITERATURE SEARCH

A systematic literature review was conducted on Ovid MEDLINE (R) spanning 1946 to June Week 1 2016 and Embase Classic + Embase spanning 1947 to 2016 Week 24. A combination of keywords was utilized including "inflammation," "cancer pain," "neoplasm," "biological marker," "allele," and "genomics." Articles were limited to studies conducted in humans and reported in English.

The literature search resulted in a total of 5,480 articles that were screened by title then abstract. Two authors screened the results (LR, AF) independently and then discussed which articles should undergo full text review until an agreed upon list was generated. If an agreement could not be reached the article underwent full text review to determine its eligibility for inclusion in the study. The full texts of all remaining articles were then obtained and independently evaluated against eligibility criteria set by the same two authors. Once full text review was complete, the two authors compared inclusion lists and discussed until consensus was reached on which articles were to be included in the review. If a consensus could not be reached a third author (RM) was consulted to arbitrate. The selection process for the current review paper is illustrated in Figure 1.

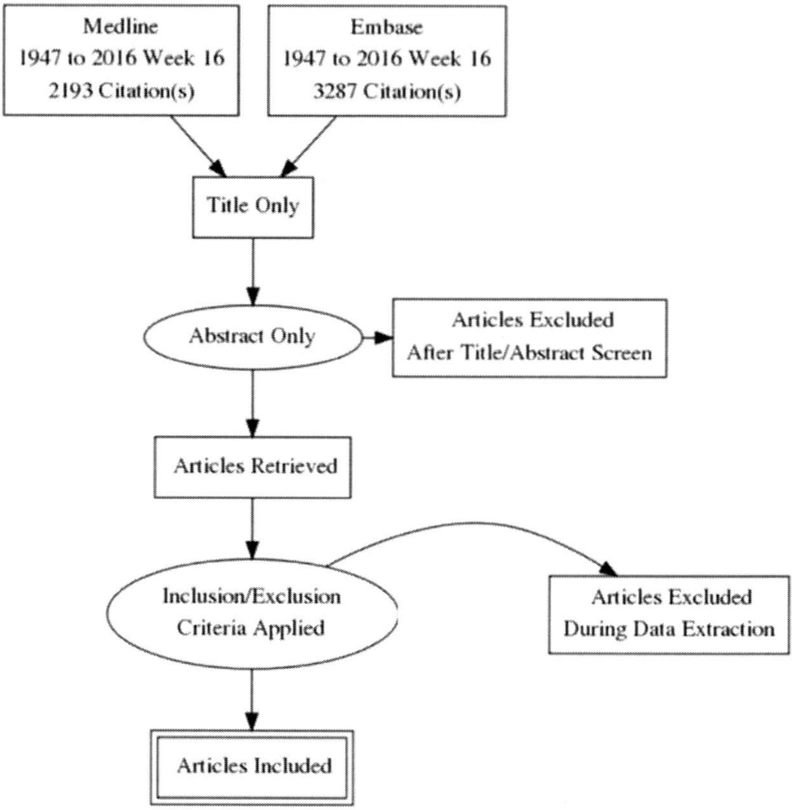

Figure 1. PRISMA flow diagram of screening process.

Screening criteria

Eligibility criteria were set forth by two authors (LR, RM) which was agreed upon by a third author (AF) prior to screening of the literature search results.

Eligibility criteria

1. The article discusses the following information
   1.1. Biological or genetic biomarkers of cancer-related symptoms or quality of life (QOL)

2. Abstracts were included in the review if
    2.1. Full text was not available
    2.2. Required information was able to be extracted from the abstract

Ineligibility criteria

1. Articles that do not involve the following information
    1.1. Biomarkers and/or their relation to cancer-symptoms or QOL
    1.2. Cancer patients or cancer survivors
2. Articles that are
    2.1. Description of a unique case or series of cases with less than 10 participants
    2.2. Systematic reviews or meta-analyses
3. Information of interest was not able to be extracted, including the following
    3.1. Information regarding the biomarker, patient outcomes and characteristics, or sample collection used

Data extraction

Two authors (LR, MC) independently extracted information of interest from full text articles or abstracts, which were included in the present review. Information of interest included characteristics of the biomarker examined, method of analyzing the biomarker, study population characteristics, main findings, and conclusions.

OUR FINDINGS

A total of 41 articles were included in this select review of the literature. The majority of articles investigated a variety of symptoms and quality of life parameters (n=20); 11 articles explored biomarkers associated with cancer-related fatigue, four articles examined biomarkers pertaining to anxiety, depressive symptoms and other cognitive complaints, and six articles investigated biomarkers of cancer cachexia. A summary of the studies can be found in Tables 1 through 4.

Table 1. Biomarkers related to fatigue (n=11)

Author (Year)	Study Information	Biomarker (Gene or Product)	Variability	Results	Ref
Bower et al. (2003)	T-cell homeostasis in breast cancer survivors with persistent fatigue (n=20) compared to non-fatigued breast cancer survivors (n=19)	Soluble IL1R1 (sIL1R1)	N/A	Fatigued survivors had increased circulating T lymphocytes, and pronounced elevation in CD4+ T lymphocytes, and CD56+ effector T lymphocytes Increased sIL1R1 levels were significantly correlated with increased circulating T cells	(9)
Bower et al. (2009)	Inflammatory biomarkers and fatigue during radiation therapy for breast (n=20) and prostate cancer (n=28)	C-reactive protein (CRP) IL1R1	N/A	CRP and IL1R1 correlated positively with fatigue	(5)
Collado-Hidalgo et al. (2006)	Plasma inflammatory markers, leukocyte subsets, and pro-inflammatory cytokines were compared between fatigued (n=50) and non-fatigued (n=50) breast cancer survivors	IL6 TNF IL1R1 sIL6R	N/A	Fatigued breast cancer survivors show increased ex-vivo monocyte production of IL6, TNF following LPS stimulation, IL1R1, and sIL6, and decreased monocyte cell-surface IL6R, activated T lymphocytes and myeloid dendritic cells Multivariate analysis revealed that the ratio of sIL6R to monocyte IL6R and also decreased circulating CD69+ T lymphocytes were highly diagnostic of fatigue ($p=0.0005$)	(10)

Author (Year)	Study Information	Biomarker (Gene or Product)	Variability	Results	Ref
Courtier et al. (2013)	Evaluate biomarkers of fatigue in early-stage breast cancer patients undergoing radiotherapy (n=100)	sIL6R	N/A	Elevated sIL6R in fatigued group compared to non-fatigued group during radiotherapy, however anxiety before treatment was strongest unique predictor for subsequent fatigue	(11)
Kurz et al. (2012)	Examine fatigue and quality of life (QOL) deterioration in lung cancer patients (n=50) and its correlation to tryptophan breakdown and immune activation markers	Trp	N/A	Patients with severe fatigue are associated with higher concentrations of inflammatory markers, increased Trp breakdown, and lower hemoglobin scores. These patients also had worse QOL and worse ECOG scores.	(13)
Light et al. (2013)	Compared leukocyte gene expression profiles in fatigued prostate cancer patients treated with androgen-deprivation therapy (n=30) to patients with chronic fatigue syndrome (CFS, n=39) with 22 controls.	*NCR31 TLR4* *CXCR3* *SOD2* *ADRB2* *VIPR2* *TNF* *LTA* *ASIC3* *P2RX7* *P2RX7* *HSPA2*	N/A	Prostate cancer (PCF) patients showed higher expression of these genes compared to control or CFS PCF patients show decreased expression of these genes compared to control or CFS CFS patients show higher P2RX7 and lower HSPA2 compared to control and PCF	(15)

Table 1. (Continued)

Author (Year)	Study Information	Biomarker (Gene or Product)	Variability	Results	Ref
Mitsunaga et al. (2010)	Examined factors correlated to serum IL-6 levels in 50 advanced pancreatic cancer patients before and after chemotherapy (1 month and 3 months later)	IL6	N/A	IL6 levels were associated with CRP levels, fatigue, and hypoalbuminemia before chemotherapy	(12)
Paddison et al. (2009)	Hematological variables correlated with fatigue in stage IIIb and IV (NSCLC) patients (n=44)	Hemoglobin Neutrophil conc. Monocyte conc.	N/A	Intense fatigue is associated with lower hemoglobin concentration Increased neutrophils and monocytes are associated with more severe fatigue	(14)
Perry et al. (2015)	Physiological factors influencing fatigue in breast cancer patients undergoing breast radiotherapy (n=24)	Caspase-1 p17 CRP	N/A	In a subset of patients, fatigue was correlated with increased caspase-1 p17 and CRP	(6)
Pertl et al. (2013)	Examined inflammatory cytokines and their associations to cancer-related fatigue and depression in breast cancer patients before chemotherapy (n=61)	CRP IFNγ IL6 TNF	N/A	Increased CRP was significantly correlated with fatigue and depression before chemotherapy	(7)

Author (Year)	Study Information	Biomarker (Gene or Product)	Variability	Results	Ref
Reinertsen et al. (2011)	Associations between single nucleotide polymorphism (SNPs) of inflammation-related genes and fatigue in 236 breast cancer survivors were analyzed	*IL1B* *IL6* *IL6R* *CRP*	rs16944 rs1800795 rs4129267 rs5855617 rs2228145 rs2794521 rs3091244	None of the SNPs or their associated mRNA levels were associated with fatigue	(8)

Table 2. Biomarkers related of cancer cachexia (n=6)

Author (Year)	Study Information	Biomarker (Gene or Product)	Variability	Results	Procedure Used	Ref
Penafuerte et al. (2016)	Aimed to identify biomarkers of cachexia in 122 newly diagnosed cancer patients (7 cancer types) categorized as having cachexia, pre-cachexia (high CRP levels), or no cachexia	Angiotensin II (AngII) TGFβ1 IL6 IL8 CRP Neutrophil-derived proteases (NDP)	N/A	Cachectic and pre-cachectic individuals show increased mRNA expression of NDP, elevated AngII, TGFβ1, and CRP ($p<0.05$) Significant increases in IL6 ($p=0.005$), IL8 ($p=0.001$), and absolute neutrophil counts ($p=0.007$) seen in cachectic patients	22 cytokine and mRNA levels isolated from blood	(16)

Table 2. (Continued)

Author (Year)	Study Information	Biomarker (Gene or Product)	Variability	Results	Procedure Used	Ref
DeJong et al. (2005)	Markers of systemic inflammation, hypermetabolism, and expression of ubiquitin and 2 uncoupling proteins were compared between weight-losing pancreatic cancer patients (n=15) and controls (n=11)	Ubiquitin IL6 sTNFR CRP UCP2 UCP3	N/A	Weight loss was correlated with systemic inflammation and increased mRNA expression of ubiquitin	Rectus abdominis muscle samples	(17)
Burney et al. (2012)	Testosterone levels in male cancer patients with (n=45) or without cachexia (n=50) were compared with non-cancer control patients (n=45)	Total testosterone (TT) and bioavailable testosterone CRP IL6	N/A	TT was significantly lower in cachexia patients compared to non-cachexia patients ($p<0.05$)	Blood samples	(18)
Op den Kamp et al. (2012)	Assessed biomarkers of the NF-κB and UPS activation pathways in NSCLC pre-cachexia patients (n=16) with <10% weight loss and compared these to 10 healthy controls	CRP sTNFR1 fibrinogen albumin	N/A	Cancer patients showed increased CRP, sTNFR1, fibrinogen, and decreased albumin compared to controls Cancer patients had significantly decreased peak oxygen consumption compared to controls	Vastus lateralis muscle biopsies	(19)

Author (Year)	Study Information	Biomarker (Gene or Product)	Variability	Results	Procedure Used	Ref
Scheede-Bergdahl et al. (2012)	Examined plasma levels of pro-inflammatory cytokines in advanced cancer patients (n=83) and their correlation to cancer cachexia	IL1B IL6 IL8 TNF	N/A	IL1B was a better predictor of cachexia characteristics than IL6 but this was not statistically significant	Blood and symptom questionnaires	(20)
Fortunati et al. (2007)	Compared serum pro-inflammatory cytokines and gene expression, and oxidative stress factors between cachectic (n=13) and non-cachectic (n=17) lung cancer patients, and also 15 matched healthy controls	TNF ROS GSH vitamin E	N/A	Cachectic patients showed significantly increased serum TNF compared to non-cachectic patients	Blood	(21)

Table 3. Biomarkers associated with depressive/anxiety/neurocognitive complaints (n=4)

Author (Year)	Study Information	Biomarker (Gene or Product)	Variability	Results	Procedure Used	Ref
Dooley et al. (2016)	Examination of the Val66Met polymorphism in *brain-derived neurotrophic factor (BDNF)* and its association to depressive symptoms in 112 early-stage breast cancer patients who recently completed cancer treatment	*BDNF*	Val66Met	In carriers of the Met allele, increased CRP levels were predictive of cognitive depressive symptoms ($p=0.04$), but not for Val/Val homozygotes	Blood and assessment questionnaires	(22)
Ng et al. (2015)	Multi-centered, prospective, longitudinal study over 3 years analyzing pro-inflammatory biomarkers and their associations to anxiety in early-stage breast cancer patients (n=101)	IL10 IL4	N/A	Increased IL10 levels ($p=0.022$) and decreased IL4 levels ($p=0.043$) were associated with increased anxiety	Blood collected prior to, during, and following chemotherapy	(23)

Author (Year)	Study Information	Biomarker (Gene or Product)	Variability	Results	Procedure Used	Ref
Patel et al. (2015)	Examined correlation between pro-inflammatory cytokines and neurocognitive dysfunction in 174 postmenopausal patients with newly diagnosed breast cancer prior to surgery	IL6 IL1R1 sTNFR2	N/A	Higher sTNFR2 was associated with worse memory performance in cancer patients, but not in the control group	Neuropsychological evaluation and blood	(24)
Archer et al. (2012)	Prospective study examining correlation between inflammatory biomarker levels and high depressive symptoms in cancer patients (56 head and neck, 34 colorectal) following surgery. Anxiety and depression were assessed prior to surgery as well as 6, 12, and 24 weeks post-surgery	CRP TNF	N/A	Baseline CRP levels were associated with depression at 6 (p=0.008) and 12 weeks (p=0.038) post-surgery in colorectal cancer patients, and baseline and 6-week TNF levels were significantly associated with increased depressive symptoms at all later time points after adjusting for clinical variables. However, these associations were not significant after adjusting for childhood trauma.	Blood prior to surgery, and 1 and 6 weeks following surgery	(25)

Table 4. Biomarkers related to multiple symptom burden and/or quality of life (n=20)

Author (Year)	Study Information	Biomarker (Gene or Product)	Variability	Results	Procedure Used	Ref
Kwekkeboom et al. (2015)	Examined the association between neuroendocrine biomarkers and the pain, fatigue, and sleep disturbance symptom cluster in advanced cancer patients receiving chemotherapy (n=14)	Salivary cortisol Salivary alpha amylase	N/A	No statistically significant correlations between symptom cluster severity and salivary cortisol or alpha amylase	Saliva 2 days before chemotherapy and symptom questionnaires	(41)
Shi et al. (2015)	Examined if any association exists between any regulatory single-nucleotide polymorphisms (SNPs) and high symptom burden in cancer patients one year following diagnosis of multiple myeloma (MM) (n=344)	*IL6* *IL1B* *TNF* *IL10*	174G>C, GG 511 CC 308G>A 1082G>A	*IL1B*-511 CC genotype associated with higher overall symptom burden in non-Hispanic whites (OR=2.35, p=0.004) *IL6*-174 GG predicted less moderate/severe fatigue in non-Hispanic whites (OR=0.53, p=0.013) but moderate/severe pain in other patients (OR=3.36, p=0.010)	Buccal-swab DNA samples	(26)

Author (Year)	Study Information	Biomarker (Gene or Product)	Variability	Results	Procedure Used	Ref
Lengacher et al. (2015)	Examined the relationship between stress hormones, cytokines, and symptoms in stage 0-III breast cancer survivors (n=322)	IL6 Cortisol	N/A	Statistically significant relationships were found between IL6 and pain ($p=0.02$) and QOL ($p<0.05$) Cortisol was significantly correlated to physical health ($p=0.05$)	Saliva samples and QOL questionnaires	(27)
Imayama et al. (2013)	Assessed associations between BMI and CRP and cancer-related symptoms and health-related quality of life (HRQOL) in stage 0-IIIA breast cancer survivors (n=483)	CRP	N/A	Increased CRP (tertile) was significantly associated with worse symptom scores including urinary incontinence, chest wall symptoms, physical functioning, role-physical, and vitality	Blood	(32)
Nguyen et al. (2015)	Examined symptom burden of individuals with chronic myelogenous leukemia (CML) and systemic mastocytosis (MS) to patients with classic myelofibrosis neoplasms (MPNs) (n=495)	*JAK2*	V617F	V617F was associated with 3 MPN symptoms (pruritus, abdominal pain/discomfort, and bone pain)	Saliva	(34)

Table 4. (Continued)

Author (Year)	Study Information	Biomarker (Gene or Product)	Variability	Results	Procedure Used	Ref
Wang et al. (2011)	Examined associations between oxaliplatin-based chemotherapy and debilitating symptoms, such as fatigue, pain, sleep disturbance, and distress in newly diagnosed colorectal cancer patients (n=31)	IL6 sTNFR1	N/A	sTNFR1 was significantly, positively associated with severity of pain, distress, drowsiness, numbness and all interference items ($p<0.05$) Increased IL6 was positively associated with disturbed sleep, distress, and mood interference ($p<0.01$)	Blood	(28)
Steel et al. (2010)	Determined prevalence and distribution of pain, fatigue, and symptoms of depression and their covariation as a cluster, and characterized how variation in each symptom and/or their covariation as a cluster was associated with changes in immunity, and determine if related to biomarkers in 206 hepatobiliary cancer patients	Eosinophil	N/A	Increased eosinophil percentage was associated with pain, fatigue, depression and increased survival	Blood	(42)

Author (Year)	Study Information	Biomarker (Gene or Product)	Variability	Results	Procedure Used	Ref
Wang et al. (2015)	Examined association between maintenance therapy and circulating inflammatory markers and patient-reported symptom burden 3-9 months after autologous stem cell transplant (AuSCT) in 51 patients with MM rated symptom severity	IL6 TNF	N/A	Elevated IL6 related to more-severe numbness/tingling Elevated TNF predictive of high-symptom severity (for symptoms including fatigue, pain, numbness/tingling, bone aches, muscle weakness)	Blood	(29)
Low et al. (2014)	Studied the association between heightened level of inflammatory markers and experience of common neurovegetative symptoms in 64 peritoneal cancer patients receiving aggressive surgical treatment	IL6 CRP	N/A	Higher levels associated with more severe fatigue, and neurovegetative symptoms of depression, poor physical QOL Higher levels associated with severe fatigue, and QOL	Blood	(30)

Table 4. (Continued)

Author (Year)	Study Information	Biomarker (Gene or Product)	Variability	Results	Procedure Used	Ref
Alexander et al. (2014)	Evaluated variations in pro- and anti-inflammatory cytokine genes between latent classes of individuals with distinct QOL trajectories in 168 patients with prostate, breast, brain, or lung cancer and 85 of their family caregivers	*IL1R2* *NFKB2*	rs4141134 rs12772374	1 or 2 of the C allele was associated with increased odds of having higher QOL 2 of G allele was associated with increased odds of having poorer QOL	Blood	(35)
Reyes-Gibby et al. (2013)	Assessed if panel of immune-response genes were associated with co-occurrence of severe pain, depressed mood, and fatigue and identified patients with severe and non-severe clusters in 599 newly diagnosed NSCLC patients	*ENOS* *IL1B* *TNFR2* *PTGS2* *IL10RB*	rs1800783 rs1143627 rs1061622 rs5275 rs2834167	These variants were predictive for symptom clusters (severe pain, fatigue, and depression)	Blood	(38)

Author (Year)	Study Information	Biomarker (Gene or Product)	Variability	Results	Procedure Used	Ref
Reyes-Gibby et al. (2013)	Examined whether variants of 37 inflammation genes may be biological markers of risk for severe pain, depressed mood, and fatigue in non-Hispanic white patients with NSCLC (n=599)	*IL8*	T251A	Most relevant genetic factor for pain, depression, and fatigue for advanced stage disease	Blood	(37)
		IL10R	Lys47Glu	Early-stage women with Lys_Glu or Glu-Glu genotype had 0.49x lower risk of fatigue compared with Lys_Lys genotype		
		IL1A	C889T	Men with C/T or T/T had lower risk of severe fatigue compared with C/C genotype		
Kao et al. (2013)	Examined the relationship between HRQOL and inflammatory markers, and the prognostic role of HRQOL in 63 malignant pleural mesothelioma (MPM) patients	CRP Vascular endothelial growth factor (VEGF) Neutrophil-to-lymphocyte ratio	N/A	Higher baseline levels of all three markers is associated anorexia, fatigue, symptomatic distress, reduced normal activity, QOL, and survival	Blood	(33)

Table 4. (Continued)

Author (Year)	Study Information	Biomarker (Gene or Product)	Variability	Results	Procedure Used	Ref
Rausch et al. (2012)	Aimed to identify SNPs related to pain in 1149 Caucasian lung cancer survivors	*LTA* *PTGS2*	rs1799964 rs5277 rs5275	Carriers reported lower pain scores Associated with worse pain, social function, and mental health Higher pain scores, lower social function scores, lower mental health scores	Blood	(40)
Raush et a.l (2010)	Aimed to identify SNPs related to symptom burden and QOL outcomes in 1149 Caucasian lung cancer survivors	*IL10* *IL1B* *IL1RN* *IL6* *IL8* *TNF*	7 SNPs 5 SNPs 7 SNPs 9 SNPs 2 SNPs 7 SNPs	21 SNPs in cytokine genes were associated with symptom burden and QOL outcomes	Blood	(39)
Bower et al. (2011)	Examine the association between inflammatory processes and fatigue, depression, and sleep disturbance in 103 female early-stage breast cancer	sTNFR2	N/A	Women treated with chemotherapy had higher levels of symptoms (fatigue, sleep problems and depressive symptoms) and higher sTNF2 than those that did not	Blood	(36)

Author (Year)	Study Information	Biomarker (Gene or Product)	Variability	Results	Procedure Used	Ref
	patients who have recently finished primary treatment (i.e., surgery, radiation, chemotherapy)			Fatigue was positively associated with sTNFR2		
Bernhard et al. (2010)	Examined the prognostic value of QOL relative to tumour marker carbohydrate antigen (CA) 19-9 and the role of CA 19-9 in estimating palliation in patients with advanced pancreatic cancer receiving chemotherapy (n=311)	CA 19-9	N/A	Patients with baseline CA 19-9 below median had marginally less pain on treatment Patients with better health status with most improvement in QOL w/in 20 days before they experienced a maximum CA 19-9 drop	Blood	(44)
Wang et al. (2010)	Assessed the association between elevations in cancer treatment-induced circulating inflammatory cytokines and development of significant symptom burden during	sTNFR1	N/A	During chemoradiation, increase in sTNFR1 was significantly related to increase in mean score for all 15 MDASI symptoms and larger radiation dose to normal lung volume	Blood	(31)

Table 4. (Continued)

Author (Year)	Study Information	Biomarker (Gene or Product)	Variability	Results	Procedure Used	Ref
	concurrent chemoradiation therapy in 62 NSCLC patients	IL6		Significantly related to increased mean severity for 5 most severe symptoms (pain, fatigue, disturbed sleep, lack of appetite, sore throat)		
Jones et al. (2008)	Examined the association between an objective measure of physical functioning, QOL, markers of systemic inflammation, and established predictors of prognosis in 42 patients with confirmed inoperable NSCLC and performance status (PS) greater than or equal to 70.	Macrophage inflammatory protein-1a	N/A	Weight loss at diagnosis is positively associated with increased macrophage inflammatory protein-1a and TNF	Blood	(43)

Author (Year)	Study Information	Biomarker (Gene or Product)	Variability	Results	Procedure Used	Ref
Mills et al. (2005)	Studied the relationship between circulating levels of soluble intracellular adhesion molecule-1 (sICAM1),VEGF, IL6, and fatigue, depressed mood, and QOL prior to and following anthracycline-based chemotherapy in 29 female stage I-IIIA breast cancer	sICAM1 VEGF IL6	N/A	sICAM1 and VEGF are related to increased fatigue and/or poorer QOL	Blood	(45)

Fatigue

Eleven articles examined possible biomarkers of fatigue (see Table 1). A commonly studied biomarker was C-reactive protein (CRP) (5–8). In the studies by Bower et al. and Pertl et al. (5, 7), CRP in the blood was found to be positively correlated with fatigue in breast and prostate cancer patients. Perry et al. (6) also found that fatigue was correlated with increased CRP levels, but only in a subset of breast cancer patients (n=24). However, when two single nucleotide polymorphisms (SNPs) of *CRP* were analyzed, no associations were found between the SNPs and fatigue, or between their associated mRNA levels and fatigue (8).

The same study by Bower et al. observed that IL1R1 was also positively correlated with fatigue in prostate and breast cancer patients during radiotherapy (5). Other studies by Bower et al. and Collado-Hidalgo et al. found that fatigued breast cancer survivors had increased circulating levels of T lymphocytes compared to non-fatigued survivors, which was also correlated to increased soluble IL1R1 levels (9, 10). In multivariate analysis in the study by Collado-Hidalgo et al. the ratio of soluble IL6R to monocyte IL6R and levels of circulating CD69+ T lymphocytes remained significantly correlated to fatigue (p=0.0005) (10).

Studies by Courtier et al. (11) and Mitsunaga et al. (12) analyzed the efficacy of IL6R and IL6 as a biomarker for fatigue in early-stage breast cancer patients undergoing radiotherapy (n=100) and pancreatic cancer patients undergoing chemotherapy (n=50), respectively (11, 12). Courtier et al. found that fatigued patients during radiotherapy had elevated soluble IL6R levels, but anxiety was the most significant predictor for fatigue, while Mitsunaga et al. found that IL6 was associated with fatigue prior to chemotherapy. However, a study by Reinertsen et al. which analyzed genetic variants of IL6, IL6R, and IL1B could not find any significant associations between the variants and fatigue in 236 breast cancer survivors (8).

In a study of 50 lung cancer patients, Kurz et al. found that increased tryptophan breakdown and decreased hemoglobin was correlated with severe fatigue (13). Paddison et al. also found that lower hemoglobin levels were positively associated with fatigue. The study also found that severe fatigue was associated with increased neutrophil and monocyte concentrations in non-small cell lung cancer (NSCLC) patients (n=44) (14). A study by Light et al. comparing gene expression profiles in fatigued prostate cancer patients (n=30) and chronic fatigue syndrome (CFS) patients (n=39) found that despite having similar symptoms, prostate cancer patients showed higher expression of certain genes (*NCR31, TLR4, CXCR3, SOD2*) and decreased expression of others (*ADRB2, VIPR2, TNF, LTA, ASIC3, P2RX7*) when compared to CFS patients (15).

Cancer cachexia

Six studies analyzed biomarkers of cachexia in cancer patients (see Table 2). Penafuerte et al. examined 122 cancer patients including seven cancer types and found that cachectic patients had significantly increased IL6, IL8, and absolute neutrophil count levels (16). In addition, both cachectic and pre-cachectic patients were observed to have significant increases in CRP, angiotensin II and TGFβ1 levels, as well as increased mRNA expression of neutrophil-derived proteases (16). Studies by DeJong et al. and Burney et al. found that cancer patients with cachexia, in comparison to cancer patients without cachexia, had increased mRNA expression of ubiquitin and significantly lower total testosterone levels, respectively (17, 18). In the study by Op den Kamp et al. pre-cachectic NSCLC patients (n=16) exhibited increased CRP, soluble TNFR1, and albumin, as well as decreased peak oxygen consumption compared to healthy controls (n=10) (19). Scheede-Bergdahl et al. examined pro-inflammatory cytokine levels (IL1B, IL6, IL8, TNF) in 83 advanced cancer patients, but did not find statistically significant associations with cancer cachexia (20). On the other hand, the study by Fortunati et al. examining lung cancer patients (13 cachectic and 17 non-cachectic) observed significantly increased TNF levels in cachectic compared to non-cachectic patients (21).

Depression, anxiety, and neurocognitive complaints

A total of four studies investigating biomarkers of depression, anxiety, or neurocognitive complaints were included in the present review. The study by Dooley et al. of 112 early-stage breast cancer patients observed that in carriers of the Met allele in the Val66Met variant of *BDNF*, increased CRP levels was a significant predictor of depressive symptoms (p=0.04), but this was not seen in Val/Val homozygotes (22). Another study of early-stage breast cancer patients (n=101) found that increased IL10 and decreased IL4 levels were significantly associated with increased anxiety (23). In a study conducted by Patel et al. a significant association was observed between soluble TNFR2 and memory performance in 174 newly-diagnosed breast cancer patients (24). Finally, the study by Archer et al. examined CRP and TNF levels in head and neck cancer patients (n=56) and colorectal cancer patients (n=34) prior to and following surgery, but found no associations to depressive symptoms after adjusting for childhood trauma (25).

Symptom burden and quality of life

20 studies examined biomarkers of symptom clusters and/or quality of life. Circulating IL6 levels, as well as genetic variants of the *IL-6* gene were analyzed in multiple studies. Shi et al. studied 344 multiple myeloma patients and found that non-Hispanic white patients with the 174GG genotype had significantly less moderate or severe fatigue, while this genotype in other patients predicted less moderate to severe pain (26). A study by Lengacher et al. revealed that IL6 levels were significantly associated to pain and QOL in 322 breast cancer survivors (27) Similarly investigating IL6 levels, Wang et al. found associations with disturbed sleep, distress, and mood interference in a study of 31 colorectal cancer patients (28). Another study by Wang et al. observed that elevated IL6 levels were associated with increased numbness or tingling in multiple myeloma patients (n=51) following stem cell transplants (29). Additional studies by Low et al. and Wang et al. observed that IL6 was associated with fatigue, depression and poor physical QOL in peritoneal cancer patients (n=64), and pain, fatigue, disturbed sleep, lack of appetite and sore throat in NSCLC patients receiving radiotherapy (n=62), respectively (30, 31).

CRP levels were also examined as biomarkers in three studies. Studies by both Imayama et al. (30) and Low et al. (32) of breast cancer survivors (n=483) and peritoneal cancer patients (n=64) found that elevated CRP was associated with decreased HRQOL (30, 32). A study by Kao et al. (33) on 63 malignant pleural mesothelioma patients found that higher baseline values of CRP, vascular endothelial growth factor (VEGF), and ratio of neutrophil to lymphocytes was associated with anorexia, fatigue and poorer QOL (33).

Variants of inflammation-related genes were also found to be possible biomarkers of QOL. The 511CC genotype for the *IL1B* gene was significantly associated with higher overall symptom burden in non-Hispanic white patients with multiple myeloma (n=344) (26) and the V617F variant in *JAK2* was associated with pruritus, abdominal pain, and bone pain in myelofibrosis neoplasm patients (n=495) (34). A study by Alexander et al. (35) on 168 cancer patients found that patients with one or two copies of the C allele in an SNP of *IL1R2* were more likely to have better QOL and those with two copies of the G allele in an SNP of *NFKB2* were more likely to have poorer QOL (35). Two studies by Wang et al. (28, 31) on colorectal cancer (n=31) and NSCLC patients (n=62) found that soluble tumour necrosis factor receptor 1 (TNFR1) was significantly and positively related to the severity of several symptoms including pain, distress, drowsiness, and numbness (28, 31). It was also associated with increased dose of radiation to normal lung tissue in the NSCLC patients (31). Increased TNF levels were also predictive of higher symptom severity in multiple myeloma patients following stem cell transplant (29). A study by Bower et al. (36) on 103 breast cancer patients observed that soluble TNFR2 was positively associated with fatigue, but this was not associated with depression or sleep problems (36).

Two studies by Reyes-Gibby et al. (37, 38) on 599 NSCLC patients found that SNPs in the genes *ENOS, IL1B, TNFR2, PTGS2,* and *IL10RB* were predictive for certain symptom clusters and specifically, SNPs in *IL1A* and *IL10R* were predictive for risk of fatigue (37, 38). Rausch et al. in two studies of 1149 Caucasian lung cancer survivors found genetic variants of *LTA* were associated with lower pain score while *PTGS2* variants were associated with worse pain, social function, and mental health. Additionally, the authors identified 21 SNPs from inflammatory genes (*IL10, IL1B, IL1RN, IL6, IL8, TNF*) associated with symptom burden and QOL outcomes (39, 40).

A study of salivary cortisol and alpha amylase in 14 cancer patients found no significant correlation to symptom cluster severity (41). However, a study of 206 hepatobiliary cancer patients conducted by Steel et al. (42) found that increased eosinophil percentages were associated with pain, fatigue and depression and Jones et al. (43) observed that increased macrophage inflammatory protein-1α (MIP1α) was associated with weight loss in 42 NSCLC patients with good performance status. Bernhard et al. (44) found that baseline levels of carbohydrate antigen 19-9 (CA 19-9) was barely significantly related to pain during chemotherapy, but not overall QOL in pancreatic cancer patients (44). Associations were found between soluble ICAM1 and VEGF, and increased fatigue and poorer QOL in 29 breast cancer patients (45).

DISCUSSION

QOL is an important consideration for patients, yet very few studies have assessed biomarkers that predict QOL. One of the barriers in finding biomarkers of QOL is the difficulty of study design, as it requires active patient input in repeated surveys, rather than simply extracting available hospital data.

However, evaluation of patients' risk of developing symptoms such as pain, fatigue, nausea, and depression will not only help them become mentally prepared for these outcomes, but also provide guidance for healthcare providers and family members in preparing the appropriate care. Moreover, life expectancy has a direct correlation with QOL (46). In addition to supporting the well-being of a patient, efforts to predict and address morbidity may also have the potential to extend their overall survival.

The commonly experienced cancer symptoms and quality of life measures span modalities ranging from physical, to cognitive, and to affective. Despite inherent differences among these variable aspects of physiology and psychology, they share several common biomarkers that are implicated in predicting adverse outcomes for cancer patients in all these measures. These biomarkers include the acute-phase protein CRP as well as members of the interleukin family of cytokines and cytokine receptors.

High levels of CRP are associated with poorer results in all measured outcomes (increased fatigue, increased depression and anxiety, higher symptom burden and worse QOL, greater cachexia) across a wide range of studies spanning prostate cancer, breast cancer, head and neck cancer, colorectal cancer, peritoneal cancer, and mesothelioma (5, 7, 8, 16, 18, 25, 30, 32, 33). In the immunological pathway, levels of soluble CRP increase in response to production of IL6 from leukocytes (47). Indeed, high levels of IL6 have also been found to be associated with increased symptom burden, lower QOL, and cachexia in multiple cancer types (10–12, 16, 27–31). Tumour necrosis factor alpha (TNF) is another pro-inflammatory cytokine also implicated in adverse cancer symptoms and QOL (19, 21, 24, 25, 28, 29, 31, 36). TNF binds to its receptor TNFR to induce intracellular signaling cascades leading to a cellular response to inflammation, including activation of IL6 transcription (47). Similar to associations involving high levels of IL6, increased levels of TNF as well as its receptors TNFR2 and TNFR1 are associated with increased depression and anxiety, increased symptom burden, and lower QOL (19, 21, 24, 25, 28, 29, 31, 36).

The most commonly implicated biomarkers also widely associated with poor cancer symptoms are all integral components of the inflammation pathway. This suggests that systemic inflammation may be part of the etiology underlying symptom clusters. Investigation of the “cytokine-immunologic model of cancer symptoms” using animal models have found that administration of pro-inflammatory agents generates sickness behaviors that mirror cancer symptoms (48). This observation not only supports inflammation as a root cause of many cancer symptoms, but also suggests inflammation as a potential area of intervention to broadly improve the wellbeing of cancer patients.

One of the methods of investigation involves identification of genetic variants in inflammatory genes associated with cancer outcome. Although the preliminary results reviewed here exhibit mixed success in identifying significant SNPs, the increased accessibility of genomic technology is likely to increase the efficiency and power of population studies. Exploration of this field may enable stratification of patients into groups of varying susceptibilities to cancer symptoms, therefore allowing healthcare practitioners to better prepare for the individual needs of a patient.

Future development of genetic biomarkers combined with mechanistic investigation of how inflammation may underpin cancer symptoms may facilitate the expansion of personalized medicine to cover all aspects of patient recovery. In order to achieve this, continued research into biomarkers associated with the subjective but debilitating effects of cancer is required.

Acknowledgments

We thank the generous support of Joey and Mary Furfari Cancer Research Fund.

References

[1] Zambroski CH, Moser DK, Bhat G, Ziegler C. Impact of symptom prevalence and symptom burden on quality of life in patients with heart failure. Eur J Cardiovasc Nurs 2005;4(3):198–206.

[2] Wong E, Chow E, Zhang L, Bedard G, Lam K, Fairchild A, et al. Factors influencing health related quality of life in cancer patients with bone metastases. J Palliat Med 2013;16(8):915–21.

[3] Chou HL, Chao TY, Chen TC, Chu CM, Hsieh CH, Yao CT, et al. The relationship between inflammatory biomarkers and symptom distress in lung cancer patients undergoing chemotherapy. Cancer Nurs 2016. doi: 10.1097/NCC.00000000 00000369.

[4] Liu L, Fiorentino L, Natarajan L, Parker BA, Mills PJ, Sadler GR, et al. Pre-treatment symptom cluster in breast cancer patients is associated with worse sleep, fatigue and depression during chemotherapy. Psychooncology 2009;18(2):187–94.

[5] Bower JE, Ganz PA, Tao ML, Hu W, Belin TR, Sepah S, et al. Inflammatory biomarkers and fatigue during radiation therapy for breast and prostate cancer. Clin Cancer Res 2009;15(17):5534–40.

[6] Perry EB, Mulay SB, Kamath J, Dowsett RJ, Neuwirth J, Grady J, et al. Factors influencing fatigue in breast cancer patients undergoing breast irradiation. J Clin Oncol 2015;33(Suppl):e20690.

[7] Pertl MM, Hevey D, Boyle NT, Hughes MM, Collier S, O'Dwyer AM, et al. C-reactive protein predicts fatigue independently of depression in breast cancer patients prior to chemotherapy. Brain Behav Immun 2013;34:108–19.

[8] Reinertsen KV, Grenaker Alnæs GI, Landmark-Høyvik H, Loge JH, Wist E, Kristensen VN, et al. Fatigued breast cancer survivors and gene polymorphisms in the inflammatory pathway. Brain Behav Immun 2011;25(7):1376–83.

[9] Bower JE, Ganz PA, Aziz N, Fahey JL, Cole SW. T-cell homeostasis in breast cancer survivors with persistent fatigue. J Natl Cancer Inst 2003;95(15):1165–8.

[10] Collado-Hidalgo A, Bower JE, Ganz PA, Cole SW, Irwin MR. Inflammatory biomarkers for persistent fatigue in breast cancer survivors. Clin Cancer Res 2006;12(9):2759–66.

[11] Courtier N, Gambling T, Enright S, Barrett-Lee P, Abraham J, Mason MD. Psychological and immunological characteristics of fatigued women undergoing radiotherapy for early-stage breast cancer. Support Care Cancer 2013;21(1):173–81.

[12] Mitsunaga S, Ikeda M, Nakachi K, Kojima Y, Oono I, Shimizu S, et al. Serum interleukin-6 level is associated with hepatic inflammatory response and fatigue in advanced pancreatic cancer. Ann Oncol 2010;21:ix19.

[13] Kurz K, Fiegl M, Holzner B, Giesinger J, Pircher M, Weiss G, et al. Fatigue in patients with lung cancer is related with accelerated tryptophan breakdown. PLoS One 2012;7(5):e36956.

[14] Paddison JS, Temel JS, Fricchione GL, Pirl WF. Using the differential from complete blood counts as a biomarker of fatigue in advanced non-small-cell lung cancer: an exploratory analysis. Palliat Support Care 2009;7(2):213–7.

[15] Light KC, Agarwal N, Iacob E, White AT, Kinney AY, VanHaitsma TA, et al. Differing leukocyte gene expression profiles associated with fatigue in patients with prostate cancer versus chronic fatigue syndrome. Psychoneuroendocrinology 2013;38(12):2983–95.

[16] Penafuerte CA, Gagnon B, Sirois J, Murphy J, MacDonald N, Tremblay ML. Identification of neutrophil-derived proteases and angiotensin II as biomarkers of cancer cachexia. Br J Cancer 2016;114(6):680–7.

[17] DeJong CHC, Busquets S, Moses AGW, Schrauwen P, Ross JA, Argiles JM, et al. Systemic inflammation correlates with increased expression of skeletal muscle ubiquitin but not uncoupling proteins in cancer cachexia. Oncol Rep 2005;14(1):257–63.

[18] Burney BO, Hayes TG, Smiechowska J, Cardwell G, Papusha V, Bhargava P, et al. Low testosterone levels and increased inflammatory markers in patients with cancer and relationship with cachexia. J Clin Endocrinol Metab 2012;97(5):E700-9.

[19] Op den Kamp CM, Langen RC, Minnaard R, Kelders MC, Snepvangers FJ, Hesselink MK, et al. Pre-cachexia in patients with stages I-III non-small cell lung cancer: systemic inflammation and functional impairment without activation of skeletal muscle ubiquitin proteasome system. Lung Cancer 2012;76(1):112–7.

[20] Scheede-Bergdahl C, Watt HL, Trutschnigg B, Kilgour RD, Haggarty A, Lucar E, et al. Is IL-6 the best pro-inflammatory biomarker of clinical outcomes of cancer cachexia? Clin Nutr 2012;31(1):85–8.

[21] Fortunati N, Manti R, Birocco N, Pugliese M, Brignardello E, Ciuffreda L, et al. Pro-inflammatory cytokines and oxidative stress/antioxidant parameters characterize the bio-humoral profile of early cachexia in lung cancer patients. Oncol Rep 2007;18(6):1521–7.

[22] Dooley LN, Ganz PA, Cole SW, Crespi CM, Bower JE. Val66Met BDNF polymorphism as a vulnerability factor for inflammation-associated depressive symptoms in women with breast cancer. J Affect Disord 2016;197:43–50.

[23] Ng T, Yap SY, Shwe M, Gan YX, Cheung YT, Fan G, et al. Evaluation of pro-inflammatory cytokines and anxiety symptoms in early-stage breast cancer (ESBC) patients: A prospective, longitudinal study. J Clin Oncol 2015;33(suppl):e20742.

[24] Patel SK, Wong AL, Wong FL, Breen EC, Hurria A, Smith M, et al. Inflammatory biomarkers, comorbidity, and neurocognition in women with newly diagnosed breast cancer. J Natl Cancer Inst 2015;107(8):djv131.

[25] Archer JA, Hutchison IL, Dorudi S, Stansfeld SA, Korszun A. Interrelationship of depression, stress and inflammation in cancer patients: a preliminary study. J Affect Disord 2012;143(1–3):39–46.

[26] Shi Q, Wang XS, Li G, Shah ND, Orlowski RZ, Williams LA, et al. Racial/ethnic disparities in inflammatory gene single-nucleotide polymorphisms as predictors of a high risk for symptom burden in patients with multiple myeloma 1 year after diagnosis. Cancer 2015;121(7):1138–46.

[27] Lengacher CA, Reich R, Paterson CL, Park J. The relationship between stress hormones (cortisol), cytokines (IL-6) and symptoms among breast cancer survivors (BCS) in an MBSR(BC) randomized controlled trial (RCT). Psychooncology 2015;24:95–6.

[28] Wang X, Cleeland CS, Johnson VE, Reuben JM, Lee B, Fogelman D, et al. sTNF-R1 and other cytokines associated with symptom burden related to chemotherapy in patients with colorectal cancer. Brain Behav Immun 2011;25:S197.

[29] Wang XS, Shi Q, Williams LA, Shah ND, Mendoza TR, Cohen EN, et al. Longitudinal analysis of patient-reported symptoms post-autologous stem cell transplant and their relationship to inflammation in patients with multiple myeloma. Leuk Lymphoma 2015;56(5):1335–41.

[30] Low CA, Bovbjerg DH, Jenkins FJ, Ahrendt SA, Choudry HA, Holtzman MP, et al. Preoperative inflammatory biomarkers and neurovegetative symptoms in peritoneal carcinomatosis patients. Brain Behav Immun 2014;42:65–8.

[31] Wang XS, Shi Q, Williams LA, Mao L, Cleeland CS, Komaki RR, et al. Inflammatory cytokines are associated with the development of symptom burden in patients with NSCLC undergoing concurrent chemoradiation therapy. Brain Behav Immun 2010;24(6):968–74.

[32] Imayama I, Alfano CM, Neuhouser ML, George SM, Wilder Smith A, Baumgartner RN, et al. Weight, inflammation, cancer-related symptoms and health related quality of life among breast cancer survivors. Breast Cancer Res Treat 2013;140(1):159–76.

[33] Kao SC, Vardy J, Harvie R, Chatfield M, van Zandwijk N, Clarke S, et al. Health-related quality of life and inflammatory markers in malignant pleural mesothelioma. Support Care Cancer 2013;21(3):697–705.

[34] Nguyen H, Hinds DA, Barnholt KE, Kiefer AK, Do CB, Eriksson N, et al. Correlation of symptom assessment with genotyping analysis of saliva samples in a large cohort of myeloproliferative neoplasm patients. Blood 2015;120(21):1732.

[35] Alexander K, Cooper B, Paul SM, West C, Yates P, Kober KM, et al. Evidence of associations between cytokine gene polymorphisms and quality of life in patients with cancer and their family caregivers. Oncol Nurs Forum 2014;41(5):E267-81.

[36] Bower JE, Ganz PA, Irwin MR, Kwan L, Breen EC, Cole SW. Inflammation and behavioral symptoms after breast cancer treatment: do fatigue, depression, and sleep disturbance share a common underlying mechanism? J Clin Oncol 2011;29(26):3517–22.

[37] Reyes-Gibby CC, Wang J, Spitz M, Wu X, Yennurajalingam S, Shete S. Genetic variations in interleukin-8 and interleukin-10 are associated with pain, depressed mood, and fatigue in lung cancer patients. J Pain Symptom Manage 2013;46(2):161–72.

[38] Reyes-Gibby CC, Swartz MD, Yu X, Wu X, Yennurajalingam S, Anderson KO, et al. Symptom clusters of pain, depressed mood, and fatigue in lung cancer: assessing the role of cytokine genes. Support Care Cancer 2013;21(11):3117–25.

[39] Rausch SM, Clark MM, Patten C, Liu H, Felten S, Li Y, et al. Relationship between cytokine gene single nucleotide polymorphisms and symptom burden and quality of life in lung cancer survivors. Cancer 2010;116(17):4103–13.

[40] Rausch SM, Gonzalez BD, Clark MM, Patten C, Felten S, Liu H, et al. SNPs in PTGS2 and LTA predict pain and quality of life in long term lung cancer survivors. Lung Cancer 2012;77(1):217–23.

[41] Kwekkeboom KL, Costanzo ES, Campbell TC. Stress biomarkers in advanced cancer patients experiencing the pain, fatigue, sleep disturbance symptom cluster. J Clin Oncol 2015;33(29):suppl 5.

[42] Steel JL, Kim KH, Dew MA, Unruh ML, Antoni MH, Olek MC, et al. Cancer-related symptom clusters, eosinophils, and survival in hepatobiliary cancer: an exploratory study. J Pain Symptom Manage 2010;39(5):859–71.

[43] Jones LW, Eves ND, Mackey JR, Peddle CJ, Haykowsky M, Joy AA, et al. Systemic inflammation, cardiorespiratory fitness, and quality of life in patients with advanced non-small cell lung cancer. J Thorac Oncol 2008;3(2):194–5.

[44] Bernhard J, Dietrich D, Glimelius B, Hess V, Bodoky G, Scheithauer W, et al. Estimating prognosis and palliation based on tumour marker CA 19-9 and quality of life indicators in patients with advanced pancreatic cancer receiving chemotherapy. Br J Cancer 2010;103(9):1318–24.

[45] Mills PJ, Parker B, Dimsdale JE, Sadler GR, Ancoli-Israel S. The relationship between fatigue and quality of life and inflammation during anthracycline-based chemotherapy in breast cancer. Biol Psychol 2005;69(1):85–96.

[46] Chu PC, Wang JD, Hwang JS, Chang YY. Estimation of life expectancy and the expected years of life lost in patients with major cancers: Extrapolation of survival curves under high-censored rates. Value Heal 2008;11(7):1102–9.

[47] Mauer J, Denson JL, Brüning JC. Versatile functions for IL-6 in metabolism and cancer. Trends Immunol 2015;36(2):92–101.

[48] Cleeland CS, Bennett GJ, Dantzer R, Dougherty PM, Dunn AJ, Meyers CA, et al. Are the symptoms of cancer and cancer treatment due to a shared biologic mechanism? A cytokine-immunologic model of cancer symptoms. Cancer 2003;97(11):2919–25.

In: Oncology
Editors: A. Furfari.et al.

ISBN: 978-1-53610-608-4

Chapter 4

BIOMARKERS OF ANALGESICS IN THE MANAGEMENT OF CANCER PAIN

Adam Bobrowski[1], BSc, Arnav Agarwal[1], MD(C), Leigha Rowbottom[1], MD(C), Rachel McDonald[1], MD(C), Anthony Furfari[1], MD(C), Stephanie Chan[1], BSc(C), Pearl Zaki[1], BSc(C), Bo Angela Wan[1], MPhil, Henry Lam[1], MLS, Azar Azad, PhD[2,3,4], Ronald Chow[1], BMSc(C), Edward Chow[1], MBBS, George S Charames[3,4,5], PhD and Carlo DeAngelis[1,*], PharmD

[1]Odette Cancer Centre, Sunnybrook Health Sciences Centre, University of Toronto, Toronto, Ontario, Canada
[2]Mount Sinai Services Inc., Toronto, Ontario, Canada
[3]Pathology and Laboratory Medicine, Mount Sinai Hospital, Toronto, Ontario, Canada
[4]Laboratory Medicine and Pathobiology, University of Toronto, Toronto, Ontario, Canada
[5]Lunenfeld-Tanenbaum Research Institute, Sinai Health System, Toronto, Ontario, Canada

* Correspondence: Carlo DeAngelis, PharmD, Odette Cancer Centre, Sunnybrook Health Sciences Centre, 2500 Bayview Ave, Toronto, Ontario, M4N 3M5, Canada. E-mail: carlodeangelis@sunnybrook.ca.

Cancer pain continues to be a significant burden to patients and a major therapeutic challenge for physicians. It is well-established that inter-individual differences in responses to both opioids and non-opioids exist. Genetic variations and short nucleotide variants (SNVs) have been attributed to these differences in analgesic efficacy and toxicity, and consequently affect dosing protocols. The purpose of this chapter is to present state-of-the-art evidence regarding biomarkers significantly associated with analgesic efficacy and toxicity related to treatment for cancer pain. A literature search was conducted in Medline and Embase utilizing the keywords "cancer," "pain," "pain management," "inflammation," and "biomarkers." Articles that reported on genetic or inflammatory biomarkers associated with cancer pain and/or the clinical efficacy and toxicity of analgesics for the management of pain were considered eligible. A total of 16 articles were included in this systematic review and 52 biochemical and genetic biomarkers were studied. This review demonstrates that limitations in tailored pain therapy, notably in opioid treatment, are manifested in individual differences in analgesic effectiveness and adverse effects. The potential use of biomarkers in the management of cancer pain is an enticing reality; however, there are few instances that a single gene determines the pain phenotype. Future research should focus on large sample populations and genome-wide associated studies to clarify the contradictions in the literature and improve the standard of care.

Introduction

Up to 28% of individuals newly diagnosed with cancer may experience pain (1) with rates increasing up to 70% in those receiving active anti-cancer therapy (2) and up to 80% among those with advanced pathology (3, 4). Relief of pain is vital to maintaining quality of life (QOL) throughout the disease process, and therefore the entire spectrum of anti-neoplastic treatment, including radiation, chemotherapy, and surgery, should be augmented by adequate analgesia for suffering patients.

The World Health Organization (WHO) proposes a three-step ladder for the management of cancer pain (5). Beginning at mild pain, a staged transition from non-opioid medication in conjunction with adjuvant analgesics is intended to titrate and optimize analgesic efficacy and function (5). In the context of moderate to severe cancer pain, the WHO recommends morphine as the first-line treatment (5). This approach to analgesia is complicated by the fact that effective morphine doses may vary widely within a population of cancer patients, ranging from 10 to 1,060 mg/day (6). Furthermore, 10 to 30% of patients achieve poor analgesic effect or experience intolerable adverse reactions with opioid use, and need to switch from morphine to an alternative opioid (7).

It has been suggested that two-thirds of the variance observed in morphine dosing is genetically based (8). Clinical studies suggest that the phenotypic heterogeneity in opioid dosing needs to be addressed in order to achieve optimal therapeutic regimens. Ultimately, two fields of study can help physicians in managing pain: pharmacogenetics, the study of inherited genetic variability in drug metabolic pathways predicating individual responsiveness to drug therapy, and pharmacogenomics, an examination of the acquired and heritable differences in many genes and gene products (9). These fields can ultimately assist the physician in meeting this ethical goal.

Allelic and biochemical differences manifest in clinical complexity by translating into patient variances at all levels of drug absorption, distribution, metabolism, and signal transduction. Such inter-individual differences require better understanding in order to optimize cancer pain management and to tailor care to the patient's therapeutic needs. Therefore, the purpose of this review of the literature is to examine biomarkers, including single nucleotide variants (SNVs), associated with cancer pain and the pharmaceutical management of cancer pain.

OUR SEARCH

A comprehensive literature search was conducted in Ovid MEDLINE (R) spanning 1946 to June Week 1 2016 and Embase Classic + Embase spanning 1947 to 2016 Week 24 with limitations to human subjects and English language. The following keywords were used: "cancer," "pain," "pain management," "biomarkers," "genetics," "genomics," and "alleles."

After the removal of identifiable duplicate hits, two authors (AB, AA) independently screened by title then abstract to identify primary studies evaluating the association between biomarkers or genetic predictors and cancer pain, as well as the pharmacokinetics and pharmacodynamics of analgesics for pain management. Consensus was reached to include or exclude potentially eligible studies, with discrepancies resolved by discussion between the two reviewers, and adjudication by a third reviewer (AF) as necessary. For all eligible titles and abstracts, full text articles were retrieved for detailed review, and were independently screened in duplicate (AB, AA), with discrepancies resolved by discussion and adjudication by a third reviewer (AF) as necessary.

For all systematic reviews identified, one reviewer (AB) examined the reference list to identify included citations to potentially eligible studies. Figure 1 provides a representation of the screening process.

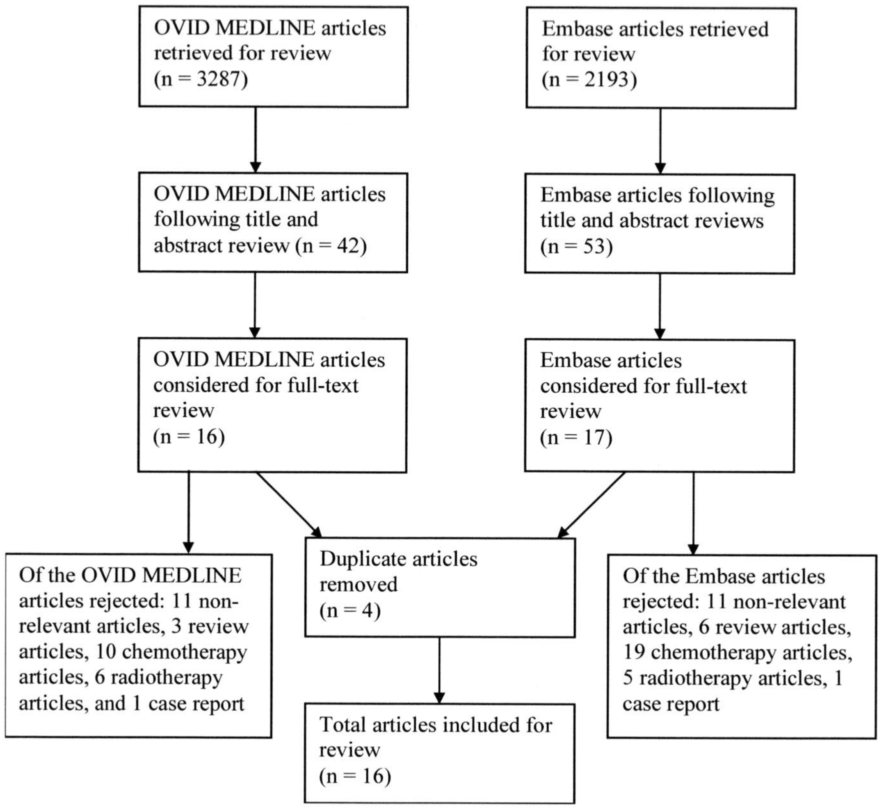

Figure 1. Flow of information diagram for publications included in systemic review.

Study eligibility

Prospective and retrospective cohort studies and clinical trials examining: a) biomarkers and genetic predictors of cancer pain; b) biomarkers and genetic predictors of opioid and other analgesic efficacy and toxicity, including bisphosphonates, and c) differences in drug receptors due to genetic variants were included in the review.

The pain evaluation method of the individual papers was not utilized as an evaluation tool for the selection of research articles.

Case reports, case series, editorials, and review articles were considered ineligible. Articles that focused primarily on chemotherapeutic agents or radiotherapy rather than analgesics were also excluded.

Data extraction

The primary outcome of the literature review was to identify biomarkers and SNVs significantly associated with analgesic response in cancer-related pain. Secondary outcomes included biomarkers and SNVs significantly associated with analgesic toxicity, and those associated with differences in patients experiencing cancer pain versus those not experiencing it. In addition to data regarding the outcomes of interest, we abstracted the main purpose, sampling and analysis methods utilized, and main general findings of each eligible study.

OUR FINDINGS

An examination of the literature yielded a total of 5,480 articles, of which 16 studies were deemed eligible and included in our qualitative analysis. Fifty-two biomarkers and SNVs were studied for their association with analgesic efficacy and toxicity in the setting of cancer pain. These markers have been summarized below and are included with the extracted article data in Table 1.

μ-Opioid Receptor (OPRM)

The *OPRM* gene encodes the G-protein coupled μ-opioid receptor, the target receptor where morphine and several other opioid analgesics exert their effects (6). Over 800 variants have been documented in the *OPRM* gene (10).

Five included studies discussed the effect of *OPRM* SNVs on the variability of opioid responsiveness, all of which studied rs1799971 (also known as A118G). SNV rs1799971 consists of an amino acid substitution from asparagine to aspartic acid in the region coding for the extracellular amino-terminus of the μ-opioid receptor. This variant has been associated with both pain and addiction, and as a part of the receptor responsible for ligand binding, may impact dose and analgesic response by an alteration in the binding affinity of different opioids (6).

Table 1. Summary of extracted biomarker and genomics data

Author	Relevant Background Information	Significant Biomarkers and Genes	Gene Variability (as applicable)	Pertinent Findings	Ref
Engler et al. 1998	Examination of bisphosphonate therapy on bone resorption parameters in cancer patients with malignant osteolytic bone disease and predictive ability of reduction in pain intensity under bisphosphonate therapy.	Urinary DPD Serum ICTP	N/A	Two bone resorption markers were significantly predictive for pain intensity reduction during bisphosphonate therapy: urinary free DPD excretion rate, and serum ICTP levels. Serum alkaline phosphatase activity also showed significant predictive value. NTx did not reach statistical significance, whereas urinary calcium excretion rate or CrossLaps excretion rate showed no effect at all. Bone AP did not show statistically significant predictive value.	(20)
Klepstad et al. 2004	Examined whether the 118 A > G variant in the *OPRM1* gene and other genetic variants in *OPRM1* influence the efficacy of morphine in cancer pain patients.	*OPRM1*	*OPRM1*: 118 A > G, -172 G > T, IVS2+31 G > A, IVS2+691 G > C	Homozygotes for 118 G allele need higher morphine doses to achieve adequate pain control relative to heterozygous and homozygous wild-type individuals ($p = 0.006$). No differences related to the -172 G > T, the IVS2+31 G > A, and the IVS2+691 G > C variants were observed.	(11)

Author	Relevant Background Information	Significant Biomarkers and Genes	Gene Variability (as applicable)	Pertinent Findings	Ref
Ross et al. 2005	Examined genetic variations in four candidate genes, the *μ-opioid receptor (OPRM1)*, *βarrestin2 (ARRB2)*, *STAT6* and *UGT2B7* genes, and their contribution to morphine response variation in cancer patients who responded to morphine versus those who switched to alternative opioids.	*OPRM1* *ARRB2* *STAT6* *UGT2B7*	*OPRM1* *ARRB2*: 1182 G/A, 5864 G/A, 8622 T/C, 11143 G/A *STAT6*: 1714 C/T, 9065 C/T *UGT2B7*: N/A	Patients who switched from morphine to other opioids were more likely to carry the common allele at 1182 G/A, 5864 G/A, 8622T/C and 11143 G/A in *ARRB2* (p = 0.021), as well as the T allele (-1714 C/T) and a different allelic frequency at 9065 C/T in the *STAT6* (p = 0.049). No genotypic or allelic differences in SNVs were seen between opioid responders and non-responders for *OPRM1* or the *UGT2B7* gene.	(6)
Martinetti et al. 2006	Evaluated the levels of bone biomarkers in patients receiving 60mg of pamidronate/week and correlating to analgesic response.	C-terminal telopeptide of type I collagen (ICTP) Osteoprotegenin (OPG)	N/A	ICTP and OPG were significantly associated with analgesic response (p = 0.001, 0.028).	(21)
Reyes-Gibby et al. 2007	Examined the joint effect of *OPRM1* and *COMT* gene variation on the clinical efficacy of morphine and in predicting morphine dose for cancer pain control.	*OPRM1* *COMT*	*OPRM1*: A118G *COMT*: Val158Met	For *COMT*, Val/Val and Val/Met genotypes required higher morphine doses compared to carriers of the Met/Met genotype.	(12)

Table 1. (Continued)

Author	Relevant Background Information	Significant Biomarkers and Genes	Gene Variability (as applicable)	Pertinent Findings	Ref
				For *OPRM1*, GG and AG genotypes required higher morphine doses compared to AA genotype carriers. In terms of joint genotype combination, *COMT* Met/Met and *OPRM1* AA genotype joint effects were found to be statistically significant for total morphine dose, median and mean pain intensity. Their joint effect results in need for less morphine dose for pain relief.	
Reyes-Gibby et al. 2008	Evaluated variants in *TNF*, *IL6* and *IL8* to explore the variability in pain and analgesic response to pain treatment with opioids.	*TNF* *IL6* *IL8*	*TNF*: -308G/A *IL6*: -174G/C, -174C/C *IL8*: -251T/A	*TNF*: -308G/A was significantly associated with pain severity at first follow-up visit (higher pain score). AA homozygote showed no difference from one visit to another; results were non-significant. Rare A allele is associated with higher TNF production. *IL6*: -174C/C genotype required 4.7 times higher dose of opioids for pain relief compared to other genotypes.	(32)

Author	Relevant Background Information	Significant Biomarkers and Genes	Gene Variability (as applicable)	Pertinent Findings	Ref
				IL6: -174G/C and *IL8* -251T/A showed improvement in pain scores. Differences in sex showed differences in pain severity in *TNF* for men and *IL6* for women. Men with *IL6* -174C/C required highest morphine dose.	
Ross et al. 2008	Examined whether functional variant in *ABCB1* is relevant to cancer patients switching from morphine to an alternative opioid, whether the *COMT* gene has influence in relation to pain and morphine dose, and identify other markers of central side-effects to morphine.	*ABCB1* *COMT*	*ABCB1*: rs2032582, rs1128503 *COMT*: rs7287550, rs174680, rs5746849, rs740603, Haplotype1, Haplotype3	*ABCB1* gene variation was associated with moderate or severe drowsiness and confusion or hallucinations. rs2032582 was associated with variability in drowsiness, confusion and hallucinations, with certain alleles resulting in decreased side-effects (p = 0.0003). *COMT* gene variation was associated independently with central side-effects. Intron 1 SNVs were associated significantly with side-effects, the most significant being rs740603 (p = 0.03). No correlations were found between morphine dose or serum morphine or metabolite concentrations, and gene variations in *ABCB1* or *COM.*	(14)

Table 1. (Continued)

Author	Relevant Background Information	Significant Biomarkers and Genes	Gene Variability (as applicable)	Pertinent Findings	Ref
Joerger et al. 2011	To assess the bone formation marker procollagen type I N-propeptide (PINP) as a biomarker for skeletal morbidity in patients with malignant bone disease.	Procollagen type I N-propeptide (PINP).	N/A	Patients with over or equal to 20% reduction in pain had lower PINP levels at baseline ($p < 0.0001$). Higher baseline PINP levels are associated with a shorter duration of pain response ($p < 0.0001$). Proportion of patients after 3 treatment cycles with ongoing pain was 18% in high baseline PINP and 45% in low PINP concentration ($p < 0.0001$).	(22)
Galvan et al. 2011	Examined new genetic variations modulating individual pain relief with the first genome-wide association study in cancer patients treated with opioids	106 genes analyzed in poor vs. good responders to opioid therapy for cancer pain; of particular interest: *SPON1, RHBDF2, ZNF235*	SNVs associated with quantitative pain relief phenotype in patients treated with opioids: rs13421094 rs12211463 rs7757130 rs2473967 rs2884129 rs7104613 rs12948783 rs10413396	Eight SNVs were strongly associated with individual response to opioid therapy for cancer pain. Strongest allelic association for quantitative pain relief phenotype was *RHBDF2* rs12948783 ($p = 8.1 \times 10^{-9}$). Genes related to nerve impulse transmission, not opioid pharmacology play the most significant role in mediating the main modulatory effect of genetic variability in pain perception. No association between pain relief and age, ethnicity or tumor type found.	(29)

Author	Relevant Background Information	Significant Biomarkers and Genes	Gene Variability (as applicable)	Pertinent Findings	Ref
Matsuoka et al. 2012	Examine genetic variations in *COMT* and *OPRM1* as predictive biomarkers in the outcome of morphine treatment.	*COMT* *OPRM1*	COMT: 472G > A OPRM1: 118A > G	Morphine treatment downregulated mRNA expression of *ARRB1* ($p = 0.003$). No correlation between *OPRM1* SNV and morphine treatment found, but A/A *COMT* genotype versus A/G and GG required a significantly lower morphine dose ($p = 0.008$).	(13)
Rakvåg et al. 2008	Examined variability in *COMT* gene regions other than rs4680 in terms of their contribution to inter-individual variability in morphine efficacy in cancer pain treatment.	*COMT*	*COMT*: rs4818, rs4680, rs5746849, rs740603, rs6269, rs2239393	Carriers of alleles for six SNVs showed weak associations with morphine dose ($p = 0.042, 0.022, 0.103, 0.099, 0.093$). Two frequent haplotypes were identified in relation to *COMT*; patients carrying the most frequent haplotype needed lower morphine doses than those not carrying the haplotype ($p = 0.005$). The alleles associated with the lowest morphine dose constituted part of the most frequent haplotype.	(17)
Heitzer et al. 2012	Examined serum cytokines correlating with pain intensity in patients with tumor-caused, advanced cancer pain.	IL7 IL18 MCP1 MIPIβ OPG	N/A	Decreases in serum concentrations of IL7, IL18, MCP1, MIPIβ and OPG were correlated with targeted pain relief with morphine ($p = 0.045, 0.016, <0.001, <0.001, <0.001$). OPG serum levels were reported to be increased in patients with nociceptive pain versus those with visceral or bone pain ($p = 0.028$ and $p = 0.019$).	(31)

Table 1. (Continued)

Author	Relevant Background Information	Significant Biomarkers and Genes	Gene Variability (as applicable)	Pertinent Findings	Ref
Oliveira et al. 2012	Study assessed effects of variants in *OPRM1*, *COMT* and *ABCB1* on cancer patient morphine consumption.	*COMT*	*COMT*: rs4680	Met allele carriers of Val158Met in *COMT* were associated with a higher dose of morphine ($p = 0.004$).	(18)
Fladvad et al. 2013	Examined whether variability in *UGT2B7, UGT1A8,* and *UGT1A1* genes contribute to variation in morphine metabolic ratios in patients with advanced cancer pain.	*UGT1A8/UGT1A1*	1a, 1d	Two haplotypes in *UGT1A8/UGT1A1* were weak predictors for reduced morphine glucuronide to morphine serum ratios after oral administration ($p < 0.1$). No association was found in the parenteral group. Clinical factors such as renal function and concomitant administration of paracetamol were also reported to be predictive to variation in serum concentration ratios ($p < 2.1{\times}10^{-12}$).	(19)
Scarpi et al. 2014	Examined clinical and genetic differences between individuals with cancer-induced bone pain (CIBP) and those with non-CIBP. Patients with CIBP were evaluated for factors contributing to opioid responsiveness.	*HTR3E* *HTR3B* (did not pass Benjamini-Hochberg criterion)	*HTR3E*: rs7432211, rs4912522 *HTR3B*: rs7103572, rs1176744	While the clinical profile of patients with CIBP and taking opioids was slightly different than that of patients without CIBP, none of the examined haplotypes were found to be significant for the association between CIBP versus non-CIBP patients, or pain relief response versus non-response.	(26)

Author	Relevant Background Information	Significant Biomarkers and Genes	Gene Variability (as applicable)	Pertinent Findings	Ref
Cajanus et al. 2016	Analyze whether SNVs in fatty acid amide hydrolase (*FAAH*) influences post-operative pain and pain sensitivities, especially cold/hot stimuli and analgesic consumption.	*FAAH*	*FAAH*: rs324420, rs1571138, rs3766246, rs4660928, rs4141964	rs324420 in *FAAH* significantly associated with lower cold pain sensitivity ($p = 0.0014$). rs1571138 associated significantly with less cold pain sensitivity ($p = 0.0014$). rs3766246 and rs4660928 nominally associated with cold pain ($p=0.0035$ and $p = 0.013$). SNV rs3766246 also nominally associated with heat pain (uncorrected $p = 0.009$). rs4141964, rs3766246, rs324420, and rs1571138 nominal association with oxycodone consumption.	(33)

A study by Klepstad et al. screened 207 Norwegian cancer patients on oral morphine titrated for pain. The authors demonstrated that those homozygous for the 118G allele of *OPRM* required greater analgesic doses to achieve pain control. No differences related to the -172 G > T, the IVS2 + 31 G > A, and the IVS2+691 G > C variants were observed (11). Similarly, Reyes-Gibby et al. (2007) genotyped 207 Caucasian cancer in-patients treated with morphine to assess the predictive ability of SNV rs1799971, among others, in morphine dosing (12). Carriers of the GG or AG genotypes needed higher morphine dose compared to their patients of the AA genotype to achieve sufficient pain relief (93% more for GG, 18% more for AG) (12). The duration of morphine treatment was also correlated to a greater morphine dose, presumably due to advanced pathology or the development of tolerance (12).

In contrast, some *OPRM* studies have yielded inconsistent results. A gene expression analysis completed within a prospective Japanese study of 41 patients with malignant neoplasms found no correlation between the *OPRM* variant and the plasma dose of morphine required to achieve analgesia (13). In addition, a study conducted by Ross et al. (2005) examined 7 SNVs (including rs1799971) related to the *OPRM* gene and found no significant differences in genotypic or allelic frequencies between morphine responders and those who had to switch to another opioid for analgesia (6).

Catechol-O-methyltransferase (COMT)

Catechol-O-methyltransferase is an enzyme involved in the metabolism of catecholamines (dopamine, epinephrine, and norepinephrine) and is encoded by the *COMT* gene. Five included studies discussed the relevance of *COMT* SNVs to pain. The majority of the studies explored rs4680 (Val158Met), an SNV that results in a substitution of methionine in place of valine and is linked to lower enzymatic activity (14). Interactions exist between the catecholamine and opioid pathways and as such, reduced COMT activity levels relate to increased catecholamine levels and also increased pain sensitivity (15, 16).

Rakvåg et al. reviewed a cohort of 197 Caucasian cancer patients, and reported that those carrying a frequent haplotype (haplotype 1) were also carriers of the A allele for the rs4680 variant and required lower morphine doses to relieve pain (17). This variant was associated with a 3 to 4-fold reduction in the activity of the COMT enzyme and an effective increase in dopamine signaling, causing a consequent reduction in levels of encephalin peptides in neurons (17). The net effect of these changes was an up-regulation of μ-opioid peptides and a compensatory alteration in pain sensitivity, where carriers of haplotype 1 required less morphine than non-carriers (17).

COMT SNVs in intron 1, particularly at position -4873G, were identified by Ross et al. (2008) to be significantly and independently associated with morphine-induced side effects in a population of 228 cancer patients experiencing moderate to severe pain (14). Although pain was cited as the primary motivation for opioid cycling, the induction of adverse events such as confusion, drowsiness, nightmares and hallucinations were important contributing factors (14).

Multi-Drug Resistance 1 (ABCB1)

The *MDR1* gene codes for P-glycoprotein (also known as ATP-binding cassette B1 (ABCB1)), which mediates the efflux of a broad range of substrates, including opioids, across the blood-brain barrier to the circulation. Decreases in the activity or concentration of P-glycoprotein have been linked to enhanced morphine analgesia, as lower levels of P-glycoprotein allow higher brain: plasma morphine levels. Genetic variation in *MDR1* has been associated with pharmacologic endpoints following morphine administration (14). The two most predominant SNVs are C3435T and C3436T. Oliveira et al. determined that some individuals homozygous for the T allele in the C3435T variant have less mRNA and consequently less protein expression, and those homozygous for the T allele in the C3436T variant have higher dosing requirements (18). Ross et al. (2008) examined 228 cancer patients utilizing morphine and found carriers of the SNV G2677T/A in exon 26 were less likely to experience morphine side effects such as drowsiness, confusion or hallucination than individuals presenting the A or T alleles at this site (14).

OPRM1, COMT and MDR1

In a study conducted by Oliveira et al. 30 cancer patients were genotyped for variants in the *OPRM1* (rs1799971), *COMT* (rs4680) and *MDR1* (rs1128503, rs1045642) genes (18). Total oral morphine equivalent consumption was only significantly correlated to carriers of the Met allele in rs4680. Individuals of this genotype required higher opioid dosing in order to achieve sufficient analgesia (18). On the other hand, Reyes-Gibby et al. (2007) showed that when compared to Met/Met homozygote cancer patients, cancer patients with the Val/Val and

Val/Met genotypes had higher morphine requirements of 63% and 23%, respectively. A joint effect between the *OPRM1* and *COMT* genes was noted, whereby patients of the *OPRM1* AA and *COMT* Met/Met genotype required the lowest amount of morphine for analgesic effect (12).

β-Arrestin 1 and 2 (ARRB1 and ARRB2)

β-arrestin is an intracellular protein involved in the down-regulation of the μ-opioid receptor (6). The absence *β*-arrestin has been associated with prolonged and increased effects of morphine (6). Ross et al. (2005) studied the genomic profile of 162 Caucasian cancer patients (117 controls and 39 opioid switchers) and compared the genotypes of patients with adequate morphine analgesia to those who switched to a different opioid (6). The study linked genetic variation in *ARRB2* with analgesic response, and demonstrated that the allele frequency for the T8622C variant was more likely to occur in opioid switchers (6). There was also a variation between switchers and controls in other common alleles, including 1182G/A, 5864G/A, and 11143G/A6 (6). In addition to these findings, a more recent Japanese study by Matsuoka et al. suggested that *β*-arrestin 1 had potential use as a biomarker for morphine therapy, finding correlation between the downregulation of *β*-arrestin1 and plasma morphine concentration ($p = 0.003$) or the required dose of morphine ($p = 0.008$) (13).

Uridine disphosphate-glucuronosyltransferase (UTG2B7)

Uridine disphosphate-glucuronosyl-transferase (UTG2B7) metabolizes morphine to morphine-3 (M3G) and morphine-6-glucuronide (M6G). M6G, an active metabolite of morphine, has significant analgesic potential and is responsible for much of morphine's pain reducing effects. A reduction in UTG2B7 activity has been associated with a reduction in morphine's analgesic potential (19). A study by Fladvad et al. observed that the genetic variability in UGT genes contributes to variability in serum morphine glucuronide to morphine concentrations in Caucasian patients (19). No effect on morphine metabolic ratios was found for variants in the *UGT2B7* gene; however, two haplotypes (1a and 1d) in *UGT1A8* and *UGT1A1* were weak predictors for reduced morphine glucuronide to morphine serum levels ($p < 0.1$) (19). Patients who did not respond to morphine were switched to oxycodone as a second-line opioid in order to utilize an alternative metabolic pathway via cytochrome P450 enzymes (19). Ross et al. (2005) observed that there was no rational for this therapeutic direction based on pharmacokinetics and that there was no difference in *UGT2B7* genotype or morphine/metabolite ratios between cancer patients switching to another opioid and controls (6).

Signal transducer and activation of transcription 6 (STAT6)

A single study explored the role of STAT6, a transcription factor involved in the up-regulation of μ-opioid receptor gene expression and identified 24 validated SNVs in *STAT*, of which seven were variants in the Caucasian test population (6). In this study, cancer patients who were unable to tolerate or gain sufficient analgesia from morphine and were transferred to a different opioid were found to have increased carriage of the variant T allele at -1714 and a significant difference in the allelic frequencies at 9065C/T in intron 6 (6). Given that the variant at -1714 did not respond to a known transcription factor binding site, it has been suggested that this SNV or one in linkage to it may be important in modifying patient response to morphine (6).

Bisphosphonates

Bone metastases are a major complication in several solid tumour cancers, particularly breast and prostate cancer, where up to 70% of cases report secondary bone tumours (20). Three studies discuss the role of bisphosphonates in inducing analgesia, utilizing biochemical markers to evaluate the efficacy of bisphosphonate treatment (20-22).

Engler et al. examined the effect of pamidronate therapy on bone resorption parameters in 70 cancer patients with malignant osteolytic bone disease (20). Specifically, they investigated how such resorption parameters were associated with various clinical outcomes, including analgesia. Reduced serum levels of C-telopeptide of type I collagen (ICTP) and urinary free deoxypyridinoline (DYD) were significant predictors reduced pain intensity, while total alkaline phosphatase (ALP) activity was associated with increased pain intensity (20). Martinetti et al. surveyed blood and urine bone markers of 42 cancer patients with bone metastases being treated with 60mg of intravenous pamidronate and found a statistically significant correlation between analgesia and a decrease in ICTP and osterprotegenin (OPG) (21). Joerger et al. assessed the bone formation marker procollagen 1 type N-propeptide (PINP) as a biomarker for skeletal

morbidity in 70 patients with malignant bone disease being treated with 60mg of intravenous pamidronate. Patients with low baseline PINP averaged a 20% or higher reduction in pain, while higher baseline PINP levels were associated with a shorter duration of pain response (22). The proportion of patients with ongoing pain following 3 bisphosphonate treatment cycles was 18% in a high baseline PINP concentration population and 45% in a low baseline PINP concentration
cohort (22).

DISCUSSION

The potential use of biomarkers for the management of cancer pain is an enticing reality that will advance the standard of analgesic care for the benefit of both patients and physicians. The major limitation for individualized pain treatment, particularly as applied to opioids, and even more specifically to morphine, manifest as differences in side effects and analgesic effectiveness among individuals. The clinical data presented in this literature review has highlighted the variability in drug responsiveness that is based on changes at the molecular level of drug target receptors, drug metabolizing enzymes, and drug transporting proteins that interact to define an altered homeostatic state.

Biomarkers have the potential of being an invaluable tool in the determination of appropriate dosing protocols. Indeed, the development of an appropriate opioid treatment schedule is a crucial and challenging step in optimizing pain control. Opioid titration is the incremental adjustment of an opioid in order to attain the optimal dose, defined as the amount of drug necessary to improve function or cause a pain reduction of at least 30% without causing adverse effect (23). Dose titration intervals are heavily dependent on opioid formulation, method of administration, absorption rates, and time to achieve a steady-state concentration (24). Nonetheless, the Fraser Institute recommends that analgesic effectiveness should not be made more frequently than every 24 hours (5 half-lives) during the titration process (25). It can be argued that the use of SNVs and biochemical signals can streamline this process and inform clinicians to administer the right dose for the right person at the right time. In this respect, the patient would experience the greatest benefit at a lower dose, limiting sedation, overdose and other unnecessary deleterious outcomes.

It is not surprising that *OPRM* studies have been inconsistent given the fact that this gene presents with 800 variants (10). However, the most prominent of these, designated rs1799971 (also known as A118G) has been found on several occasions to impact analgesic dose and addictive potential (6). Similarly, a review of *COMT* showed that patients carrying the Met/Met genotype required lower morphine doses than the Val/Val genotype group with the rs4680 SNV (Val158Met variant) cohort being intermediate between these two extremes. Genetic variation in the β-arrestin 2 gene (*ARRB2*) correlated with a switch from morphine to an alternative opioid; absence of the β-arrestin 2 intracellular protein was found to be associated with reduced levels of the μ-opioid receptor (6). On the other hand, the *UGT2B7* variation did not prove to alter morphine/metabolite ratios between cancer patients that switched opioid medications (6). Patients found to carry the variant T allele at -1714 and a difference in allelic frequencies of 9065C/T in intron 6 of *STAT6* was unable to gain sufficient analgesia from morphine (6). A study by Scarpi et al. examined 2,201 Caucasian patients with clinical and genetic differences between individuals with cancer-induced bone pain (CIBP) and those with non-CIBP. While the clinical profile of patients with CIBP and taking opioids was slightly different than that of patients without CIBP, none of the examined haplotypes were found to be significant for the association between CIBP versus non-CIBP patients, or pain relief response versus non-response (26). Interestingly, patients with low baseline PINP averaged a 20% or higher reduction in bone pain, while the proportion of patients with ongoing pain following three bisphosphonate treatment cycles was reported to be 45% in a similar low baseline PINP concentration cohort, in comparison to only 18% in the high baseline PINP population (22).

It is evident that there are few instances where pain phenotypes are determined by a single gene. Codeine, an opioid analgesic that exerts its analgesic effect primarily through its metabolism to morphine (5-15%) by *CYP2D6*, is a case in point. CYP2D6, primarily expressed in the liver, is responsible for the oxidative metabolism of a variety of drugs including tamoxifen, opioids, and antipsychotics (27). There are more than 80 allele variations of the *CYP2D6* gene, each with different effects and variable phenotypic functional impact. *CYP2D6* pharmacogenetics has defined 5-10% of people poor metabolizers, 2-11% intermediate metabolizers, 77-92% extensive metabolizers and 1-2% ultra-rapid metabolizers, based on enzymatic activity. Morphine concentrations in extensive metabolizers and ultra-rapid metabolizers are highly variable and several studies have documented either a lack of analgesic effect in poor metabolizers or life-threatening side effects in ultra-rapid metabolizers including morphine toxicity in breastfed infants of ultra-fast metabolizing mothers (28).

Added to these disparities is the fact that there may be individual genetic variants that do not have a significant effect on their own but that may participate in additive, synergistic, or antagonistic relationships with other genetic variants, ultimately producing clinically significant outcomes. Therefore, it is likely that further study of the interactions and relationships between multiple genes will be more revealing than a review of single independent gene associations.

The situation becomes even more complicated given that Galvan et al. monitored one million SNVs in 1008 cancer patients from 11 European countries and identified eight SNVs associated with opioid analgesia. Remarkably, these SNVs were linked with genes involved in processes related to nerve physiology rather than the pharmacology of opioids. Such genes included *RHBDF2* (rs12948783), *SPON1* (rs7104613), and *ZNF235* (rs10413396), and may represent novel targets for the advent of alternative analgesic therapeutics (29).

Systemic inflammation is a further complication that is associated with tumours and has been shown to suppress hepatic metabolism; for example, *CYP3A4* was shown to be responsible for the breakdown of the majority of anticancer drugs. Tumour derived cytokines and elevated inflammatory markers are a significant feature of the metabolic imbalance underlying cachexia and an exacerbation of the pain response. Drug transporters such as breast cancer resistance protein (BCRP) and multidrug resistance-associated protein 2 (MRP2) are altered in kidney as well as liver indicating that the bio-distribution of drugs in several organs may be affected in the presence of cancer (30). Nonetheless, the utilization of measurable inflammatory biomarkers could be an invaluable tool for evaluating pain intensity. A study by Heitzer et al. evaluated 19 biomarkers in a population of 20 healthy patients experiencing no pain and 20 cancer patients with exacerbated pain (31). Utilizing changes in a numeric rating scale (NRS) to determine changes in pain scores before and after morphine treatment, 6 biomarkers, including IL7, IL8, MCP1, MIP1α, MIP1β, and OPG, were found to be significantly correlated with pain relief (31). Reyes-Gibby et al. (2008) examined the influence of cytokine gene variants in predicting pain severity in relation to morphine uptake in 140 lung cancer patients and found the variants -308G/A in the gene for *tumour necrosis factor-α* (*TNF*) and -174G/C in the gene for *Interleukin 6* (*IL6*) to be significantly associated with pain severity and morphine equivalent daily dose respectively (32). Indeed, patients with the *IL6* -174CC genotype required 4.7x higher doses of opioids for analgesia compared to carriers homozygous and heterozygous for the G allele (32). Although there is an abundance of evidence related to the impact of ethnicity on innate immune genetic markers related to cancer treatments, majority of studies used in this review did not separate or use ethnicity as criteria in their analysis, whereas some focused on specific ethnicity and excluded others (19). All of these factors predicate targeting specific cytokines for safer and more effective use of both chemotherapeutic and analgesic agents.

Adjuvants add to this discussion in line with the WHO model (5). A recent study by Cajanas et al. explored the influence of SNVs in the gene coding for fatty acid amide hydrolase (*FAAH*) on cold (n = 900) and heat (n = 1,000) pain sensitivity and oxycodone consumption in Finnish breast cancer patients undergoing surgery. A missense mutation (rs324420) was the most likely variant in reducing FAAH activity and was significantly associated with lowered cold pain sensitivity and nominally associated with postoperative oxycodone consumption. An additional SNV was found to be significantly associated with cold pain intensity (rs1571138), with other SNVs showing nominal cold pain association and reduced postsurgical opioid consumption. By way of explanation, N-arachidonyl ethanolamide (AEA), the natural substrate of FAAH, is believed to interact with the opioid system to enhance analgesic effect (33). In a similar manner, morphine and tetrahydrocannabinol (THC) potentiate one another in terms of analgesic effect and can be inhibited by either a cannabinoid or an opioid receptor antagonist (34). The net effect of these interventions is more effective analgesia at a lower dose of opioid drug with fewer side effects. From this standpoint, FAAH and the cannabinoids remain the targets of further investigation both from the point of view of genetic mechanism and therapeutic application.

Although these studies contribute to the knowledge base, there are several contradictions in the literature. Further inquiry and research is required to develop an evidence-based pain management protocol that considers both biochemical and genetic signatures of a patient. Many of the reviewed publications analyzed only a few genes and suffered inconsistencies due to a small sample population. A comprehensive study of the biomarkers outlined herein would require large sample sizes, significant statistical power, and the employment of gene-wide associated studies. The direction of future research should explore these techniques and current and novel biomarkers with the objective of establishing an individualized analgesic treatment protocol for cancer pain.

Acknowledgments

We thank the generous support of Joey and Mary Furfari Cancer Research Fund.

References

[1] Vuorinen E. Pain as an early symptom in cancer. Clin J Pain 1993;9(4):272–8.

[2] Portenoy RK, Miransky J, Thaler HT, Hornung J, Bianchi C, Cibas-Kong I, et al. Pain in ambulatory patients with lung or colon cancer. Prevalence, characteristics, and effect. Cancer 1992;70(6):1616–24.

[3] Donnelly S, Walsh D. The symptoms of advanced cancer. Semin Oncol 1995;22(2 Suppl 3):67–72.

[4] Caraceni A, Portenoy RK. An international survey of cancer pain characteristics and syndromes. IASP Task Force on Cancer Pain. International Association for the Study of Pain. Pain 1999;82(3):263–74.

[5] World Health Organization. Cancer pain relief: with a guide to opioid availability 1996. URL: http://apps.who.int/iris/bitstream/10665/37896/1/9241544821.pdf.

[6] Ross JR, Rutter D, Welsh K, Joel SP, Goller K, Wells AU, et al. Clinical response to morphine in cancer patients and genetic variation in candidate genes. Pharmacogenomics J 2005;5(5):324–36.

[7] Cherny N, Ripamonti C, Pereira J, Davis C, Fallon M, McQuay H, et al. Strategies to manage the adverse effects of oral morphine: an evidence-based report. J Clin Oncol 2001;19(9):2542-54.

[8] Riley J, Ross JR, Rutter D, Shah S, Gwilliam B, Wells AU, et al. A retrospective study of the association between haematological and biochemical parameters and morphine intolerance in patients with cancer pain. Palliat Med 2004;18(1):19-24.

[9] Ito RK, Demers LM. Pharmacogenomics and pharmacogenetics: future role of molecular diagnostics in the clinical diagnostic laboratory. Clin Chem 2004; 50(9):1526-7.

[10] Riva A, Kohane IS. A SNP-centric database for the investigation of the human genome. BMC bioinformatics 2004;5(1):33.

[11] Klepstad P, Rakvåg TT, Kaasa S, Holthe M, Dale O, Borchgrevink PC, et al. The 118 A> G polymorphism in the human μ-opioid receptor gene may increase morphine requirements in patients with pain caused by malignant disease. Acta Anaesthesiol Scand 2004;48(10):1232-9.

[12] Reyes-Gibby CC, Shete S, Rakvåg T, Bhat SV, Skorpen F, Bruera E, et al. Exploring joint effects of genes and the clinical efficacy of morphine for cancer pain: OPRM1 and COMT gene. Pain 2007;130(1):25-30.

[13] Matsuoka H, Arao T, Makimura C, Takeda M, Kiyota H, Tsurutani J, et al. Expression changes in arrestin β 1 and genetic variation in catechol-O-methyltransferase are biomarkers for the response to morphine treatment in cancer patients. Oncol Rep 2012;27(5):1393.

[14] Ross JR, Riley J, Taegetmeyer AB, Sato H, Gretton S, du Bois RM, et al. Genetic variation and response to morphine in cancer patients. Cancer 2008;112(6):1390-403.
[15] Zubieta JK, Heitzeg MM, Smith YR, Bueller JA, Xu K, Xu Y, et al. COMT val158met genotype affects μ-opioid neurotransmitter responses to a pain stressor. Science 2003;299(5610):1240-3.
[16] Nackley AG, Tan KS, Fecho K, Flood P, Diatchenko L, Maixner W. Catechol-O-methyltransferase inhibition increases pain sensitivity through activation of both β 2- and β 3-adrenergic receptors. Pain 2007;128(3):199-208.
[17] Rakvåg TT, Ross JR, Sato H, Skorpen F, Kaasa S, Klepstad P. Genetic variation in the catechol-O-methyltransferase (COMT) gene and morphine requirements in cancer patients with pain. Mol Pain 2008;4:64.
[18] Oliveira A, Dinis-Oliveira R, Nogueira A, Azevedo A, Goncalves F, Silva P, et al. 975 COMT genetic variation may influence opioid dosing requirements in the treatment of cancer-related pain. Eur J Cancer 2012;48(Suppl 5):S235.
[19] Fladvad T, Klepstad P, Langaas M, Dale O, Kaasa S, Caraceni A, et al. Variability in UDP-glucuronosyltransferase genes and morphine metabolism: observations from a cross-sectional multicenter study in advanced cancer patients with pain. Pharmacogenet Genomics 2013;23(3):117-26.
[20] Engler H, Koeberle D, Thuerlimann B, Senn HJ, Riesen WF. Diagnostic and prognostic value of biochemical markers in malignant bone disease: a prospective study on the effect of bisphosphonate on pain intensity and progression of malignant bone disease. Clin Chem Lab Med 1998;36(11):879-85.
[21] Martinetti A, Ripamonti C, Miceli R, Seregni E, Mariani L, De Conno F, et al. Short-term effects of pamidronate on bone turnover: Can bone markers be considered predictive of the analgesic response?. Oncol Rep 2007;17(6):1533-40.
[22] Joerger M, Templeton A, Köberle D, Engler H, Riesen WF, Thürlimann B. Procollagen type I N-propeptide is a predictor of skeletal morbidity in patients with malignant osteolytic bone disease on bisphosphonates. Cancer Chemother Pharmacol 2011;67(5):1137-44.
[23] McMaster University Michael G DeGroote National Pain Centre. Canadian guideline for opioid use for pain, 2016. URL: http://nationalpaincentre.mcmaster.ca/opioid/cgop_b02_r09.html.
[24] Portenoy RK. Treatment of cancer pain. Lancet 2011;377(9784):2236-47.
[25] Fraser Health. Hospice Palliative Care Program Symptom Guidelines: Principles of opioid management, 2006. URL: https://www.fraserhealth.ca/media/16FHSymptomGuidelinesOpioid.pdf.
[26] Scarpi E, Calistri D, Klepstad P, Kaasa S, Skorpen F, Habberstad R, et al. Clinical and genetic factors related to cancer-induced bone pain and bone pain relief. Oncologist 2014;19(12):1276-83.
[27] Samer CF, Lorenzini KI, Rollason V, Daali Y, Desmueles JA. Applications of CYP450 testing in the clinical setting. Mol Diagn Ther 2013;17(3):165-184.
[28] Mulugeta L, US Food and Drug Administration. Pharmacogenetics of codeine. URL: http://www.fda.gov/downloads/advisorycommittees/committeesmeetingmaterials/pediatricadvisorycommittee/ucm343582.pdf.

[29] Galvan A, Skorpen F, Klepstad P, Knudsen AK, Fladvad T, Falvella FS, et al. Multiple loci modulate opioid therapy response for cancer pain. Clin Cancer Res 2011; 17(13):4581-7.

[30] Robertson G, Liddle C, Clarke S. Inflammation and altered drug clearance in cancer: transcriptional repression of a human CYP3A4 transgene in tumor-bearing mice. Clin Pharmacol Ther 2008;83(6):894–7.

[31] Heitzer E, Sandner-Kiesling A, Schippinger W, Stohscheer I, Osprian I, Bitsche S, et al. IL-7, IL-18, MCP-1, MIP1-β, and OPG as biomarkers for pain treatment response in patients with cancer. Pain Physician 2012;15(6):499-510.

[32] Reyes-Gibby CC, Osta BE, Spitz MR, Parsons H, Kurzrock R, Wu X, et al. The influence of tumor necrosis factor-alpha -308 G/A and IL-6 -174 G/C on pain and analgesia response in lung cancer patients receiving supportive care. Cancer Epidemiol Biomarkers Prev 2008;17(11):3262–7.

[33] Cajanus K, Holmström EJ, Wessman M, Anttila V, Kaunisto MA, Kalso E. Effect of endocannabinoid degradation on pain: role of FAAH polymorphisms in experimental and postoperative pain in women treated for breast cancer. Pain 2016;157(2):361-9.

[34] Manzanares J, Julian MD, Carrascosa A. Role of the cannabinoid system in pain control and therapeutic implications for the management of acute and chronic pain episodes. Curr Neuropharmacol 2006;4(3):239-57.

In: Oncology
Editors: A. Furfari et al.

ISBN: 978-1-53610-608-4

Chapter 5

CHEMOPREVENTATIVE CAPACITY OF NON-STEROIDAL ANTI-INFLAMMATORY DRUGS (NSAIDS)

Adam Bobrowski[1], BSc, Arnav Agarwal[1], MD(C), Anthony Furfari[1], MD(C), Rachel McDonald[1], MD(C), Leigha Rowbottom[1], MD(C), Stephanie Chan[1], BSc(C), Pearl Zaki[1], BSc(C), Bo Angela Wan[1], MPhil, Henry Lam[1], MLS, Azar Azad[2,3,4], PhD, Ronald Chow[1], BMSc(C), George S Charames[3,4,5], PhD, Edward Chow[1], MBBS and Carlo DeAngelis[1,*], PharmD

[1]Odette Cancer Centre, Sunnybrook Health Sciences Centre, University of Toronto, Toronto, Ontario, Canada
[2]Mount Sinai Services Inc., Toronto, Ontario, Canada
[3]Pathology and Laboratory Medicine, Mount Sinai Hospital, Toronto, Ontario, Canada
[4]Laboratory Medicine and Pathobiology, University of Toronto, Toronto, Ontario, Canada
[5]Lunenfeld-Tanenbaum Research Institute, Sinai Health System, Toronto, Ontario, Canada

[*] Correspondence: Carlo DeAngelis PharmD, Odette Cancer Centre, Sunnybrook Health Sciences Centre, 2075 Bayview Ave, Toronto, Ontario, M4N 3M5, Canada. E-mail: carlodeangelis@sunnybrook.ca.

Inflammation has been found to be an important part of the carcinogenic process. As such, potential use of non-steroidal anti-inflammatory drugs (NSAIDs) to lower inflammation and mitigate the risk of carcinogenesis has been investigated. Genetic variations and short nucleotide variants (SNVs) may contribute to differences in the efficacy of NSAIDs in cancer prevention. The purpose of this chapter is to present evidence regarding biomarkers significantly associated with the chemopreventative capacity of regular NSAID use. A literature search was conducted in Medline and Embase databases utilizing the keywords "cancer", "inflammation" and "biomarkers". Articles that reported on genetic or protein biomarkers associated with cancer prevention, cancer recurrence, and/or the clinical efficacy and toxicity of NSAIDs for cancer treatment were considered eligible. Eight articles were included in this review and 33 genetic predictors were found to be associated with the prophylactic efficacy and toxicity of NSAID use in cancer. The current literature related to colorectal, lung, and prostate cancers has failed to consistently support the ability of NSAIDs to reduce cancer risk; however, a negative association between breast cancer and the use of COX-2 (also known as PTGS2) inhibitors has been suggested. The inconsistent nature of the current state of research demands further study and exploration into the pharmacogenetics of NSAIDs.

INTRODUCTION

The connection between cancer and inflammation was identified as early as 1863 when German physician Rudolf Virchow (1821-1902) observed the presence of leukocytes in cancerous tissue (1). Initially, the presence of immune cells within the majority of neoplastic microenvironments was considered a natural attempt of the host's immune system to eradicate the pathology. Although the immune system initiates an anti-tumoural response, a mounting volume of evidence over the past decade demonstrates that chronic inflammation, initiated by innate immune cells designated to combat disease and heal wounds, plays a paradoxical role in aiding the progression of neoplastic lesions. Inflammation acts as a point of access for bioactive molecules, including growth factors, survival factors and angiogenic factors. This acquisition allows neoplasms to sustain cellular proliferation, evade death signals, undergo epithelial-mesenchymal transition (EMT) activation, and modify the extracellular matrix allowing for angiogenesis, invasion and eventual metastasis (2). Additionally, the generation of reactive oxygen and nitrogen species by cancer-related inflammation contributes to neoplasia development through causing genomic instability, ultimately leading to DNA mutations (3-5). The updated 2011 version of the seminal article "The hallmarks of cancer" added inflammation as a hallmark of the disease (2).

Supporting the central role inflammation plays in cancer pathology, the regular use of non-steroidal anti-inflammatory drugs (NSAIDs) has been associated with reducing cancer and recurrence risk. NSAIDs provide antipyretic, anti-thrombotic, analgesic, and anti-inflammatory effects through the inhibition of two isoforms of the cyclooxygenase enzyme, the catalyst of the rate-limiting step in the conversion of arachidonic acid to prostaglandins (6). Cyclooxygenase 1 (COX-1), also known as prostaglandin-endoperoxide synthase 1 (PTGS1) is constitutively expressed in many cell types and assumes multiple physiologic roles. Cyclooxygenase-2 (COX-2), also known as prostaglandin-endoperoxide synthase 2 (PTGS2) is a pro-inflammatory enzyme that is induced by mitogens, oncogenes, growth factors, hormones and cytokines (6). Prostaglandin H2 (PGH2), the product of PTGS1, is converted by prostaglandin E2 synthase to prostaglandin E2 (PTE2) which stimulates cancer progression. Additionally, the peroxidase activity of PTGS2 can lead to an increase of oxidative stress and the overexpression of this enzyme in breast cancer has been associated with an increase in chromosomal aberrations (7).

The prophylactic efficacy of NSAIDs has been evaluated in chemopreventative tests, including the Prevention of Colorectal Sporadic Adenomatous Polyps (PreSAP) and Adenoma Prevention Celecoxib (APC) trials. The APC trial demonstrated that in comparison to placebo, PTGS2 specific inhibitor celecoxib at both low (200mg) and high doses (400mg) reduced adenoma size and number, as well as the risk of recurrence (33 and 45% respectively) in patients with a history of colorectal cancer. The PreSAP trial reached a similar conclusion as a 400mg dose of daily celecoxib reduced colorectal adenoma recurrence by 36% within 3 years following polypectomy (8, 9). Although these studies reflect the benefit of NSAID use in reducing cancer risk, the complexity manifests in inter-individual allelic and biochemical differences that impact NSAID efficacy. These differences need to be better understood to minimize toxicity and the risk of cancer initiation and reoccurrence. Genetic differences in inflammation-related pathways, particularly in the formation of prostaglandins, can assist in pinpointing who will benefit most from NSAID therapy and minimize associated adverse effects.

Therefore, the purpose of this chapter is to examine biomarkers, including single nucleotide variants (SNVs), associated with the protective role of NSAIDs in cancer prevention and reoccurrence.

Our search

A comprehensive literature search was conducted in Ovid MEDLINE (R) spanning 1946 to June Week 1 2016 and Embase Classic + Embase spanning 1947 to 2016 Week 24 with limitations to human subject and English language. The following specified search terms were employed: "cancer", "biomarkers", "genetics", "genomics", "alleles" and "inflammation".

Titles and abstracts of all articles were screened for relevancy independently by two authors (AA, AB). Any discrepancies were resolved by discussion amongst the two authors, with adjudication by a third reviewer (AF) if necessary. If the articles were thought to be relevant based on the eligibility criteria by both authors, full-text publications were retrieved and examined for relevancy. This process was completed independently by two authors (AA, AB) who then resolved any discrepancies to reach a final list of selected articles. If consensus could not be reached, a third author (AF) was consulted. For all systematic reviews identified, one reviewer (AB) examined the reference list to identify included citations that were potentially eligible studies. The main purpose, sampling and analysis methods utilized, and main general findings of each eligible study were extracted.

Study eligibility

All prospective and retrospective cohort studies and clinical trials that examined the following topics were included: a) biomarkers and genetic predictors of efficacy of NSAIDs, b) cancer prevention or prophylaxis and c) differences in drug receptors or response due to gene variants. Publications were excluded if they were review articles, case reports, or case series.

The primary outcome of the literature review was to identify biomarkers and SNVs significantly associated with the efficacy of NSAIDs as prophylactic pharmaceuticals for cancer prevention.

Our findings

A review of the literature yielded a total of 5,480 articles, of which 63 were initially deemed eligible based on title and abstract review. Following further full-text screening, 8 were deemed eligible and were included in this review (see Figure 1). Thirty-three biomarkers and SNVs were associated with variability in the efficacy of NSAIDs for the prophylaxis of carcinogenesis. These disease-related biomarkers are summarized below and in Table 1.

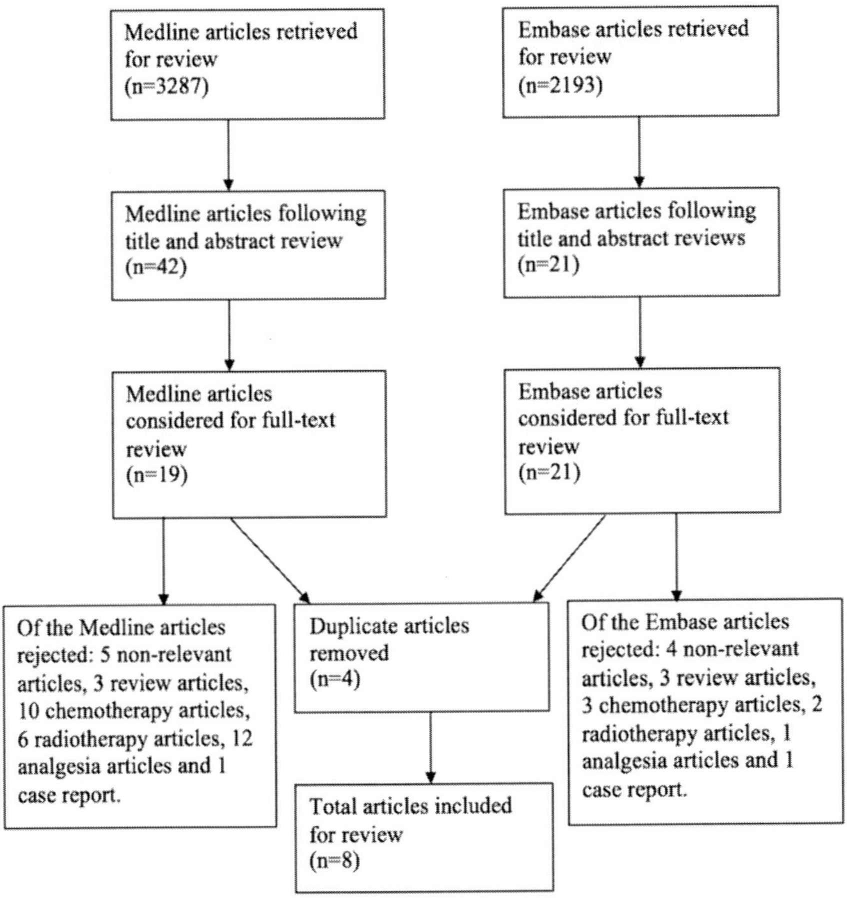

Figure 1. Flow of information diagram for publications included in systematic review.

Table 1. Data extracted from full-text articles

Author	Study Aim	Biomarker	Gene Variability	Result	Procedure Used	Ref
Barry et al., 2009	Investigate whether polymorphisms in *PTGS2* influenced risk of colorectal adenoma recurrence among 979 participants in the Aspirin/Folate Polyp Prevention Study.	*PTGS2*	*PTGS2*: rs5277, rs4648310	Evaluation of interaction between any *PTGS2* genotype and aspirin treatment on risk of adenoma recurrence: • rs5277: Homozygous carriers of the minor C allele had a 51% increase in risk over the GG homozygotes. • rs4648310: Heterozygous AG carriers had a 37% increased risk compared to AA homozygotes. • Consumption of 81mg of aspirin treatment daily was associated with a protective effect among AA homozygotes.	Genomic DNA was isolated from whole blood cells with proteinase K digestion and phenol-chloroform extraction and *PTGS2* polymorphisms were genotypes with the 5'nuclease TaqMan allelic discrimination assay using the ABI 7900 Sequence Detection System.	(13)
Habermann et al., 2014	Explore the interactions between genes and NSAID use in relation to colorectal cancer risk, as well as how genetic variants interact with inflammation markers C-reactive protein (CRP) and serum amyloid A (SAA).	*PTGS2* *FLT1* *NRP1* CRP SAA	*PTGS2*: rs2383529 *FLT1*: rs1886233 *NRP1*: rs7895812	• *NRP1* SNP is associated with colorectal cancer risk. • There was a protective association between NSAID use and PTGS2 SNP for colorectal cancer risk • FLT1 SNP was associated with a decreased colorectal cancer risk. • Genetic variations altered serum concentrations of CRP and SAA	SNPs were genotyped using Illumina GoldenGate and Sequenom platforms. Latex-enhanced nephelometry used to quantify CRP and SAA	(14)
Kopp et al., 2013	Evaluate whether polymorphisms and haplotypes in inflammation-related genes are associated with the risk of prostate cancer	*NFKB1* *PTGS2*	*NFKB1*: -94ins/delATTG *PTGS2*: COX-2 +8473T->C	• *NFKB1*: Nonusers of NSAIDs with deletion had a lowered risk of PC. • NSAID use was associated with aggressive PC, but not risk.	Genomic DNA extracted from blood samples. DNA obtained from frozen leukocytes.	(18)

Author	Study Aim	Biomarker	Gene Variability	Result	Procedure Used	Ref
	(PC), prostate cancer aggressiveness and the interactions between polymorphisms and NSAID use.			• Both polymorphisms showed an additive effect with BMI where obese carriers of the wild type of both polymorphisms exhibited a reduced PC risk, whereas all men showed reduce risk with either polymorphism.		
Kraus et al., 2013	Examine whether genetic polymorphisms in several inflammation-related genes modify the effects of celecoxib on recurrence in patients with adenomatous polyps, and whether these polymorphisms are related to cardiovascular and gastrointestinal toxicities/symptoms.	Adenoma recurrence *PTGS1* *PTGS2* *EGFR* *ALOX12* *ALOX15* *PGDH* Gastrointestinal toxicity and symptoms: *EGFR* Cardiovascular toxicity and symptoms: *PTGS1* *EGFR* *PTGES*	Adenoma recurrence *PTGS1*: rs10306164, rs1236913, rs1330344, rs3119773, rs10306110, rs10306122 *PTGS2*: rs4648268, rs6489469, rs5275 *EGFR*: rs7801956 *ALOX12*: rs2073438, rs2292350, rs4796535 *ALOX15*: rs11568141 *PGDH*: rs7349744, rs1365613 Gastrointestinal toxicity and symptoms *EGFR*: rs2072454 Cardiovascular toxicity and symptoms: *PTGS1*: rs10306164 *EGFR*: rs4947984, rs4947986, rs3752651 *PTGES*: rs2302821, rs2241271	• Genetic variability in *PTGS1, PTGS2,* EGFR and ALOX12/15 may contribute to adenoma recurrence. • A number of genetic variants, especially PTGES, CRP, SRC and GPX3 variants, were associated with increased cardiovascular toxicity risk among patients on celecoxib. EGFR gene variants were also associated with gastrointestinal toxicity.	Whole blood samples; genotyping of lymphocyte DNA with Illumina genotyping platform.	(12)

Table 1. (Continued)

Author	Study Aim	Biomarker	Gene Variability	Result	Procedure Used	Ref
Menegaux et al., 2013	Assesse for role of NSAIDs (particularly inhibitors of *PTGS2*) in breast cancer occurrence and the role of *PTGS2* polymorphisms and their interactions with NSAIDs in breast cancer in 811 patients and 829 controls.	*PTGS2*	*PTGS2*: rs5275, rs20417, rs5277	• Overall, a negative association was found between breast cancer and NSAID use (OR = 0.83). Negative association most pronounced among women consuming selective *PTGS2* inhibitors (OR = 0.60). • Negative association of NSAID use and breast cancer found for rs5275 (OR = 0.73). • Positive association of breast cancer and NSAID use found for rs5277 (OR = 1.22). • Odds ratio between NSAID use and *PTGS2* polymorphism was 1.25 for rs20417 and 1.16 for rs5275.	Participants were surveyed for NSAID consumption and then blood samples were taken and genotyped with an Illumina chip.	(17)
Slattery et al., 2007	Evaluate whether polymorphisms in *IL6* and vitamin D receptor gene (*VDR*) and the regular use of NSAIDs/aspirin impacts risk of colon and rectal cancers.	*IL6* *VDR*	*IL6*: rs1800796, rs1800795 *VDR*: Fok1, Bsm1	• Patients with C allele in *IL6* rs1800796 polymorphism and GG genotype in the *IL6* rs1800795 polymorphism were associated with a reduced colon cancer risk (OR = 0.76), but a higher rectal cancer risk (OR = 1.49). • Patients with a C allele in either of the *IL6* polymorphisms had a reduced risk of colon cancer if currently using NSAIDs/aspirin. • Patients with a *IL6* rs1800795 (-174G->C) polymorphism using	Data retrieved from two case-control studies that used identical methods. Blood drawn for genotyping with TaqMan-based assays.	(11)

Author	Study Aim	Biomarker	Gene Variability	Result	Procedure Used	Ref
				NSAIDs/aspirin had lower risk for rectal cancer. • Patients using NSAIDs/aspirin with f allele in *VDR* Fok1 and C allele in *IL6* rs1800796 had lower colon cancer risk. • Patients using NSAIDs/aspirin with bb genotype in *VDR* Bsm1 and CC genotype in *IL6* rs1800795 had lower risk of rectal cancer. While patients with these polymorphisms and not using NSAIDs had lower colon cancer risk.		
Vogel et al., 2007	Analyzed 355 adenocarcinoma cases and 753 control cases from the Danish "Diet, Cancer and Health" study to find polymorphisms linking colorectal cancer risk with polymorphisms associated with inflammatory risk and possible interaction with lifestyle choices like duration and use of NSAIDs, smoking and alcohol consumption	*OCTN* *DGL5*	*OCTN*: C1672T *DGL5*: R30Q	• C allele carriers (WT) of *OCTN* C1672T had an increased risk of colorectal cancers only among NSAID users rather than nonusers (IRR = 1.60). While variant allele carriers who were non-users had an IRR of 1.36 and those who were NSAID users were not at increased risk of CRC. • WT allele of DGL5 R30Q was at a higher risk of colorectal cancer among NSAID users than among non-users. Variant allele carriers who were non-NSAID users had a higher risk (IRR = 1.14) of CRC than users who had a lower risk (IRR = 0.63).	Genomic DNA extracted from blood samples. DNA obtained from frozen leukocytes.	(15)

Table 1. (Continued)

Author	Study Aim	Biomarker	Gene Variability	Result	Procedure Used	Ref
Vogel et al., 2008	Analyzed 428 lung cancer cases and 800 control cases from the Danish "Diet, Cancer and Health" study to find polymorphisms linking lung cancer risk with polymorphisms associated with inflammatory risk and possible interaction with lifestyle choices like duration and use of NSAIDs, smoking and alcohol consumption	*PTGS2* *IL1B* *PPARG*	*PTGS2*: A-1195G *IL1B*: T-31C *PPARG*: Pro^{12}Ala	• *PTGS2* A-1195G and NSAID use – Non-NSAID users with variant G-allele carriers had lower risk for lung cancer (IRR = 0.54), whereas NSAID use reduced risk of lung cancer among WT carriers (IRR = 0.64), but not among variant allele carriers. • *IL1B* T-31C and NSAID use – Non-users with variant allele were at a higher risk of lung cancer than homozygous WT carriers, whereas NSAID users had 30% higher risk of lung cancer than WT nonusers of NSAIDs (not statistically significant) • *PPARG* Pro^{12}Ala and NSAID use – Among nonusers, carriers of variant allele were at lowered risk compared to homozygous WT carriers. Among NSAID users, homozygous WT carriers were at reduced risk of lung cancer but variant allele carriers had higher lung cancer risk of lung cancer.	Genomic DNA extracted from blood samples. DNA obtained from frozen leukocytes.	(16)

Colorectal cancer (CRC)

Individuals suffering from inflammatory bowel disease (IBD), including Crohn's disease and ulcerative colitis, have demonstrated a predisposition to the development of colorectal cancer (CRC). It has been postulated that this is due to the inflammatory nature of the disease, and as such, the regular use of anti-inflammatory drugs such as aspirin and other NSAIDs could mitigate this risk (10, 11). In a multi-center case-control review of colon cancer (1,579 cases and 1,977 controls) and rectal cancer (794 cases and 1,005 controls) patients, Slattery et al. reported that regular aspirin/NSAID use independently modifies the risk of CRC. The study also found that the C allele of the *IL6* rs1800796 variant and the GG genotype of *IL6* rs1800795 variant were associated with a significant reduction of colon cancer, but not rectal cancer. Individuals with a C allele in either variant who were current users of aspirin/NSAIDs had the lowest colon cancer risk (11).

The vitamin D receptor (VDR) has been designated an important mediator of many physiologic pathways, including inflammation (11). Variants in the *VDR* gene, namely of the f, t, B and S alleles, result in a less robust VDR activity and have been related to greater cell-mediated immune response (11). Although the relationship did not reach statistical significance, a dramatic decrease in CRC risk was noted in individuals with both the f allele of the rs2228570 (Fok1) variation of *VDR* and a C allele in either *IL6* variant (11). Concurrent aspirin/NSAID use by carriers of the C allele in the *IL6* rs1800796 variant and f allele in Fok1 showed a significant reduction in CRC risk ($p = 0.05$) (11). On the other hand, the Bsm1 *VDR* variant in aspirin/NSAID non-users with the rs1800795 *IL6* variant lowered the risk of CRC ($p = 0.01$) (11).

Kraus et al. genotyped 117 Israeli CRC patients in order to characterize the heritability of adenoma recurrence and celecoxib cardiotoxicity under the influence of NSAIDs. *PTGS1* (rs10306164, rs1236913, rs1330344, rs3119773), *PTGS2* (rs4648268, rs6489469), *ALOX12* (rs2073438, rs4796535) and *ALOX15* (rs11568141) genes were related to adenoma recurrence in controls versus the NSAID treatment group, while variants rs2241271 and rs2302821 of the *prostaglandin E synthase* gene (*PTGES*) were associated with a 4-and 8-fold increased risk in cardiac toxicities respectively. Three *SRC* gene variants (rs6017996, rs6018256 and rs6018257) were associated with a significantly increased risk of cardiovascular toxicity ranging from 7-to-11 fold, and two variants in *GPX3* and the rs10306164 variant in *PTGS1* also resulted in an increased risk of cardiotoxicity, with the rs1800947 *CRP* variant conferring the highest risk (OR = 15.14). On the other hand, three variants of the *IL23R* gene significantly lowered the risk of cardiovascular toxicity in the NSAID treatment group. Although selective NSAIDs such as the PTGS2 inhibitors have been reported to demonstrate improved gastrointestinal tolerability, patients taking celecoxib with the rs2255888 variant in *ALOX15* and two variants of the *EP4* gene were shown to be at increased risk of such adverse events (12).

Barry et al. examined the hypothesis that aspirin lowers the risk of colorectal adenomas and cancer by inhibiting PTGS2 activity (13). Two *PTGS2* SNVs have been associated with significant increase risk of adenoma reoccurrence (13). Carriers of the minor C allele in the rs5277 *PTGS2* variant demonstrated a 51% increased CRC risk while the AG heterozygotes of the rs4648310 *PTGS2* variant had a 37% increased risk compared to the AA homozygotes (13). The use of 81mg of aspirin daily was found to have a possible protective effect on rs4648310 *PTGS2* AA homozygotes not seen in the AG heterozygotes. It was suggested that multiple pathways, in addition to PTGS2, including PTGS1 could be involved in the mechanism for this affect. The authors concluded that although the results provided support to the hypothesis that PTGS2 plays a role in the etiology of colon cancer and might be a target for aspirin chemoprevention, the associations were modest and potentially due to chance (13).

In addition to inflammatory markers alone, Habermann et al. explored the interaction between NSAID use on CRC risk and genes related to inflammation and angiogenesis. The authors selected 821 Caucasian women aged 50-79 from the Women's Health Initiative Observational Study and found that genetic variations in angiogenic and inflammatory pathways altered this association. SNVs in angiogenesis genes, *NRP1* (rs7895812) and *FLT1* (rs1886233), were significantly associated with CRC risk ($p = 0.02$) and decreased CRC risk with NSAID use ($p = 0.02$). NSAID use had a protective effect against CRC for individuals with the wild-type for the *PTGS2* rs2383529 variant ($p = 0.01$) (14).

A study by Vogel et al. examined a sample of 355 adenoma carcinoma cases and 753 controls to determine if variants associated with altered inflammatory response correlated with CRC risk, and investigated the role of lifestyle factors such as alcohol consumption, smoking and NSAID use (15). The results could not confirm that inborn variations in the inflammatory response had any major influence on the role of CRC risk, or that any of the variants were associated with the risk of CRC (15).

Lung cancer

Vogel et al. examined the effects of NSAIDs on a study population of 428 lung cancer cases within a study cohort of 800 persons (16). The epidemiologic findings regarding the effect of NSAID use on the risk of lung cancer were conflicting. Several NSAIDs repressed PTGS2 and resulted in less PGD2 and apoptosis. However, an activator of PTGS2 known as c-MYB that is implicated in lung cancer was not modified by NSAID use. In fact, variant allele carriers of *PTGS2* A/G or *PPARG* Pro12Ala were at lower cancer risks in the absence of NSAID use, and NSAID use was only associated with lower cancer risk among double homozygous carriers. This epidemiological data suggests that NSAID use should be considered and studied when associations between genetic variants and cancer risk are evaluated, however no firm conclusions can be made at this time (16).

Breast cancer

Menegaux et al. conducted a population-based, case-controlled study conducted between 2005 and 2008 using in-person interviews, self-administered questionnaires, and blood sampling of 811 cases and 821 controls. The authors reported a negative association between NSAID use and breast cancer risk (OR = 0.83), more pronounced among women who consumed PTGS2 selective inhibitors (OR = 0.60). There was an inverse relationship between breast cancer and the absence of the common allele for *PTGS2* rs5275 (OR = 0.73) and a positive association close to statistical significance for the rare allele for *PTGS2* rs5277 (OR = 1.22). These results were interpreted to suggest that regular NSAID consumption led to lower breast cancer risk, particularly in young women using PTGS2 selective inhibitors (17).

Prostate cancer

Kopp et al. analyzed 370 prostate cancer cases in order to determine whether SNVs and NSAID use were associated with prostate cancer risk. Inflammation was not associated with aggressive prostate cancer, and consistent with this finding, there was a lack of correlation between prostate cancer risk and SNVs of *PTGS2*. NSAID use was not identified as a protective factor in prostate cancer. Interestingly, carriers of the -94ins/delATTG variant allele of *NFKB1* who did not use NSAIDs and men of a body mass index <30kg/m^2 were found to be at lower risk of this pathology (18).

Discussion

The use of biomarkers for the prophylaxis of cancer and prevention of its recurrence is an advantageous goal that will advance the standard of care in oncology for both the benefit of the patient and clinician. A disruption of tissue homeostasis and repair common to both inflammation and cancer has raised the theoretical construct that aspirin/NSAID-based compounds represent a pharmacologic solution to both pathologies. Genetic variants in inflammation-related pathways, and particularly in prostaglandin synthesis, may help identify individuals who will benefit most from NSAID therapy while minimizing associated toxicities. A review of the literature has revealed studies conducted in the context of CRC, lung cancer and prostate cancer. Currently, these studies have failed to consistently support the hypothesis of NSAID use for chemoprevention. The inconsistent nature characterizing this field requires further research and pharmacogenetics inquiry.

NSAIDs, particularly salicylates, present a complexity of interactions in and of themselves and have been deemed to work via multiple pathways, including inhibition of nuclear factor kappa B (NFκB) and its upstream activator IK kinase beta (IKBKB) (19). These proteins regulate immune cell differentiation and survival, and sensitizing insulin signaling (19). In patients with type II diabetes, treatment with aspirin has been shown to reduce fasting plasma glucose, total cholesterol, C-reactive protein (CRP), triglycerides and insulin clearance while reducing hepatic glucose production and improving insulin stimulated peripheral glucose uptake by 20% (20). Insulin and inflammatory pathways intersect, and aspirin/NSAIDs have been shown to modulate the link between Bsm1 *VDR* variant and CRC risk (21). VDR has concurrently been involved in multiple pathways associated with both inflammation and insulin. VDR has been designated an immune system regulator by increasing the production of inflammatory cytokines IL2 and IL4, while also playing a key role in vitamin D signaling, cell proliferation and differentiation. It is not surprising that vitamin D deficiency has been shown to increase the severity of inflammatory bowel disease, which carries a marked excess risk of CRC (22). Undetermined factors that modify the inflammation pathways may contribute to the different associations of colon and rectal cancer with regards to NSAID use, SNV associations, and response to aspirin's protective effect.

The potential for therapeutic efficacy of NSAIDs becomes even more complicated when considering their dose- and treatment duration-dependent cardiovascular and gastrointestinal toxicities. The APC trial found that compared to placebo, patients receiving low and high doses of celecoxib had a respective 2.6- and 3.4-fold increased risk for adverse cardiac events (8, 9). Rofecoxib is a selective PTGS2 inhibitor that was withdrawn from the market because its use over a five-year span resulted in 88,000 to 140,000 severe cardiac events. Although the mechanisms behind NSAID-induced cardiotoxicity is contentious, well-recognized cardiac effects including myocardial infarction, thrombosis, blood pressure elevation and destabilization, pulmonary edema and death have been documented (23). The efficacy and toxicity of NSAIDs are married by their ability to inhibit PTGS enzymes, and recognition of this double-edged phenomenon speaks to the difficulty of interpreting their biological impact. Kraus et al. recognized the relationship between genetic variation in several candidate pathways defining cardiac and gastric toxicities, in addition to the variable potential of NSAIDs to lower adenoma recurrence, and once again highlights the need for further study into the genetic factors impacting NSAID use (12).

Adjuvants add to this discussion because animal models have suggested that mobilization of the endogenous cannabinoid system through the inhibition or genetic deletion of enzymes responsible for the metabolism of arachidonoylethanolamide (AEA) and 2-arachidonoylglycerol (2-AG) can have beneficial effects on pain as well as NSAID-induced gastric damage (24). Gastric hemorrhages in mice models induced by 100mh/kg of diclofenac sodium were moderated by the use of FAAH inhibitors, increasing AEA levels, as well as by monoacylglycerol lipase inhibitors, increasing 2-AG, and in FAAH knockout mice (25, 26). A study by Holt et al. indicated that the amide of ibuprofen, 2-amino-3-methylpyridine (Ibu-AM5) was two to three orders of magnitude more potent than ibuprofen itself towards inhibition of FAAH, while retaining PTGS-inhibitory activity and also presenting a lower ulcerogenic effect than the parent drug (27). A study by Capriani et al. reported that the N-(3-methylpyridin-2-yl) amide derivative of flurbiprofen, Flu-AM1, was a dual-action inhibitor of FAAH and PTGS in mouse brain homogenates, with the ideal composition of FAAH:PTGS inhibition still requiring further investigation (28). These observations not only presages drugs with joint FAAH/PTGS inhibitory activity, but also explains that we need to do more work in defining complex biomedical systems.

RNA interference (RNAi), a method of silencing genes post-transcriptionally, may represent an alternative to NSAIDs as it can be used to knockdown PTGS2 expression in cancer cell lines, potentially achieving the same-antitumorigenic effects as NSAIDs without NSAID-specific cardiovascular and gastrointestinal toxicities. Using a specific siRNA to knockdown endogenous PTGS2 expression in HT-29 colon cancer cell lines, Charames et al. observed a significant decrease in PTGS2 mRNA and protein levels, as well as aberrant elevation of 15-PGDH, a physiological antagonist of PTGES (29).

Ultimately, the nascent study of biomarkers outlined herein defines the need for increased biochemical investigations, studies of SNV associations to response, and possibly gene-wide associated studies (GWAS) of significant statistical power for a comprehensive identification of more SNVs with potential associations, with the goal of establishing personalized medicine as a therapeutic goal. The literature as it applies to NSAIDs and cancer serves to raise our index of suspicion, and if we are mindful of what we do not know, we will continue to do good service to the field of oncology, and to its practitioners, and most importantly to those who suffer.

Acknowledgement

We thank the generous support of Joey and Mary Furfari Cancer Research Fund.

References

[1] Balkwill F, Mantovani A. Inflammation and cancer: back to Virchow? Lancet 2001;357(9255):539–45.

[2] Hanahan D, Weinberg RA. Hallmarks of cancer: The next generation. Cell 2011;144(5):646–74.

[3] Miwa H, Kanno H, Munakata S, Akano Y, Taniwaki M, Aozasa K. Induction of chromosomal aberrations and growth-transformation of lymphoblastoid cell lines by inhibition of reactive oxygen species-induced apoptosis with interleukin-6. Lab Invest 2000;80(5):725–34.

[4] Shacter E, Weitzman SA. Chronic inflammation and cancer. Oncology (Williston Park) 2002;16(2):217–26, 229, 230-2.

[5] Kiraly O, Gong G, Olipitz W, Muthupalani S, Engelward BP. Inflammation-induced cell proliferation potentiates DNA damage-induced mutations in vivo. PLoS Genet 2015; 11(2):e1004901.

[6] Mazhar D, Ang R, Waxman J. COX inhibitors and breast cancer. Br J Cancer 2006;94(3):346–50.

[7] Menter DG, Schilsky RL, Dubois RN. Cyclooxygenase-2 and cancer treatment: Understanding the risk should be worth the reward. Clin Cancer Res 2010;16(5):1384–90.

[8] Arber N, Eagle CJ, Spicak J, Racz I, Dite P, Hajer J, et al. Celecoxib for the prevention of colorectal adenomatous polyps. N Engl J Med 2006;355(9):885–95.

[9] Bertagnolli MM, Eagle CJ, Zauber AG, Redston M, Solomon DH, Kim K, et al. Celecoxib for the prevention of sporadic colorectal adenomas. N Engl J Med 2006; 355(9):873–4.

[10] Choi PM, Zelig MP. Similarity of colorectal cancer in Crohn's disease and ulcerative colitis: implications for carcinogenesis and prevention. Gut 1994;35(7):950–4.

[11] Slattery ML, Wolff RK, Herrick JS, Caan BJ, Potter JD. IL6 genotypes and colon and rectal cancer. Cancer Causes Control 2007;18(10):1095–105.

[12] Kraus S, Hummler S, Toriola AT, Poole EM, Scherer D, Kotzmann J, et al. Impact of genetic polymorphisms on adenoma recurrence and toxicity in a COX2 inhibitor (celecoxib) trial. Pharmacogenet Genomics 2013;23(8):428–37.

[13] Barry EL, Sansbury LB, Grau MV, Ali IU, Tsang S, Munroe DJ, et al. Cyclooxygenase-2 polymorphisms, aspirin treatment, and risk for colorectal adenoma recurrence--data from a randomized clinical trial. Cancer Epidemiol Biomarkers Prev 2009;18(10):2726–33.

[14] Habermann N, Brown E, Toth R, Scherer D, Buck K, Cheng TYD, et al. Abstract 2186: Inflammation- and angiogenesis- related genes: Interaction with NSAID use, and serum inflammatory markers on colorectal cancer risk within the Women's Health Initiative. Cancer Res 2014;74(19 Supplement):2186.

[15] Vogel U, Christensen J, Dybdahl M, Friis S, Hansen RD, Wallin H, et al. Prospective study of interaction between alcohol, NSAID use and polymorphisms in genes involved in the inflammatory response in relation to risk of colorectal cancer. Mutat Res 2007;624(1-2):88–100.

[16] Vogel U, Christensen J, Wallin H, Friis S, Nexø BA, Raaschou-Nielsen O, et al. Polymorphisms in genes involved in the inflammatory response and interaction with NSAID use or smoking in relation to lung cancer risk in a prospective study. Mutat Res 2008;639(1-2):89–100.

[17] Menegaux F, Truong T, Ajmia F, Anger A, Mulot C, Laurent-Puig P, et al. Abstract 141: Nonsteroidal anti-inflammatory drugs (NSAIDs), cyclooxygenase-2 polymorphisms, and breast cancer risk. Cancer Res 2013;73(8 Supplement):141.

[18] Kopp TI, Friis S, Christensen J, Tjønneland A, Vogel U. Polymorphisms in genes related to inflammation, NSAID use, and the risk of prostate cancer among Danish men. Cancer Genet 2013;206(7-8):266–78.

[19] Yuan M, Konstantopoulos N, Lee J, Hansen L, Li ZW, Karin M, et al. Reversal of obesity- and diet-induced insulin resistance with salicylates or targeted disruption of Ikkbeta. Science 2001;293:1673–7.

[20] Hundal RS, Petersen KF, Mayerson AB, Randhawa PS, Inzucchi S, Shoelson SE, et al. Mechanism by which high-dose aspirin improves glucose metabolism in type 2 diabetes. J Clin Invest 2002;109(10):1321–6.

[21] Slattery ML, Samowitz W, Hoffman M, Ma KN, Levin TR, Neuhausen S. Aspirin, NSAIDs, and colorectal cancer: possible involvement in an insulin-related pathway. Cancer Epidemiol Biomarkers Prev 2004;13:538–45.
[22] Fang V, Foye WO, Robinson SM, Jenkins HJ. Hypoglycemic activity and chemical structure of the salicylates. J Pharm Sci 1968;57(12):2111–6.
[23] New Scientist. Up to 140,000 heart attacks linked to Vioxx 2005. URL: https://www.newscientist.com/article/dn6918-up-to-140000-heart-attacks-linked-to-vioxx/
[24] Fowler CJ. Monoacylglycerol lipase - a target for drug development? Br J Pharmacol 2012;166(5):1568–85.
[25] Roques BP, Fournié-Zaluski M-C, Wurm M. Inhibiting the breakdown of endogenous opioids and cannabinoids to alleviate pain. Nat Rev Drug Discov 2012;11(4):292–310.
[26] Sasso O, Bertorelli R, Bandiera T, Scarpelli R, Colombano G, Armirotti A, et al. Peripheral FAAH inhibition causes profound antinociception and protects against indomethacin-induced gastric lesions. Pharmacol Res 2012;65(5):553–63.
[27] Holt S, Paylor B, Boldrup L, Alajakku K, Vandevoorde S, Sundström A, et al. Inhibition of fatty acid amide hydrolase, a key endocannabinoid metabolizing enzyme, by analogues of ibuprofen and indomethacin. Eur J Pharmacol 2007;565(1-3):26–36.
[28] Cipriano M, Björklund E, Wilson AA, Congiu C, Onnis V, Fowler CJ. Inhibition of fatty acid amide hydrolase and cyclooxygenase by the N-(3-methylpyridin-2-yl) amide derivatives of flurbiprofen and naproxen. Eur J Pharmacol 2013;720(1-3):383–90.
[29] Charames GS, Bapat B. Cyclooxygenase-2 knockdown by RNA interference in colon cancer. Int J Oncol 2006; 28(2):543-9.

In: Oncology
Editors: A. Furfari et al.
ISBN: 978-1-53610-608-4

Chapter 6

BIOMARKERS OF RADIATION RESPONSE

Matthew Choi*[1]*, BSc, Leigha Rowbottom*[1]*, MD(C), Rachel McDonald*[1]*, MD(C), Anthony Furfari*[1]*, MD(C), Stephanie Chan*[1]*, BSc(C), Bo Angela Wan*[1]*, MPhil, Pearl Zaki*[1]*, BSc(C), Azar Azad*[2,3,4]*, PhD, Ronald Chow*[1]*, BMSc(C), Carlo DeAngelis*[1]*, PharmD, Henry Lam*[1]*, MLS, Edward Chow*[1]*, MBBS and George S Charames*[3,4,5,*]*, PhD

[1]Odette Cancer Centre, Sunnybrook Health Sciences Centre, University of Toronto, Toronto, Ontario, Canada
[2]Mount Sinai Services Inc., Toronto, Ontario, Canada
[3]Pathology and Laboratory Medicine, Mount Sinai Hospital, Toronto, Ontario, Canada
[4]Laboratory Medicine and Pathobiology, University of Toronto, Toronto, Ontario, Canada
[5]Lunenfeld-Tanenbaum Research Institute, Sinai Health System, Toronto, Ontario, Canada

[*] Correspondence: George S Charames PhD, FACMG, Pathology and Laboratory Medicine, Mount Sinai Hospital, Suite 6-500, 600 University Ave, Toronto, Ontario, M5G 1X5, Canada. E-mail: George.Charames@sinaihealthsystem.ca.

Inflammation is a hallmark of radiation therapy and an increase of various pro-inflammatory cytokines, enzymes, and blood biomarkers occurs after radiation therapy. However, increased inflammation may result in variable outcomes in patients after radiation therapy. Certain inflammatory markers are associated with increased survival, decreased local recurrence and better QOL, while others are associated with poorer outcomes. Many pro-inflammatory genes are implicated in the prognosis of patients after radiation therapy. This chapter summarizes the literature that has previously investigated biomarkers of radiation response. The clinical utility of biomarkers for predicting response to treatment is significant and may assist with personalization of cancer care and appropriate resource allocation. A literature search was conducted on Medline and Embase databases using keywords including "inflammation", "cancer", "radiation", "neoplasm", "cancer", "pain", "biological marker", "allele", and "genomics". Results were limited to studies conducted in humans and English language. The search generated a total of 5,480 results that were screened by two authors (MC, AF) by title, abstract, then full text using defined criteria. A total of four studies were included. A trend amongst these studies was the increase in inflammation after radiation therapy in patients. However, the link between inflammation and the poor outcomes is not fully supported, due to contrasting findings. Genes implicated with inflammation should be further studied to examine their predictive capabilities and for genetic variants that may allow for better prognosis and treatment.

INTRODUCTION

Radiotherapy has been widely recognized as a staple in cancer treatment throughout the world. In fact, over 52% of cancer patients across a wide variety of cancer types are prescribed radiotherapy thus making it an important component of a patient's care plan (1). Radiotherapy is utilized with both curative and palliative intents, depending on the stage of cancer and site of radiation. However, delivery of radiotherapy does not result in homogenous patient response, with some patients experiencing positive responses and others receiving only limited benefit. Furthermore, radiotherapy may occasionally cause adverse events which can negatively impact patient functioning and quality of life (QOL). For radiation oncologists and patients, the decision to undergo radiotherapy can be difficult, as patients may fear they will not receive any benefit despite having to endure various side effects. This consideration is especially relevant for palliative cancer patients, where the goals of treatment are to provide symptom relief and improve QOL. It is advantageous for such patients to avoid undergoing treatment where the risk of adverse events is high and the chance of therapeutic response is low, ensuring their QOL is not compromised at the end of life.

Currently, there is large focus in cancer research towards the development of precision medicine with an emphasis on the use of genetic and protein biomarkers to enhance patient centered care. Recent studies have discovered various biomarkers which have been associated with overall survival, tumour response, tissue toxicity, and even treatment failure in radiotherapy patients (2–4).

The clinical utility of such biomarkers is enormous given the fact that only approximately 30% of patients receive complete remission from pain, while a larger proportion experience negative side effects which can compromise their QOL. Additionally, treatment delivery greatly strains the healthcare system; the cost of one session of radiation therapy can cost upwards of a hundred dollars depending on the type of radiation (5). With the prevalence of cancer increasing due to an aging population, it is imperative that treatment be cost effective. Biomarkers that predict whether or not patients will respond to radiotherapy would allow physicians to make better treatment decisions and also reduce healthcare costs.

This chapter summarizes the literature investigating biomarkers of radiation response. The biomarkers herein reported may serve as predictors of a range of different responses to radiation. Ultimately, biomarkers have the potential to assist radiation oncologists with treatment prescription and may benefit future radiotherapy patients.

OUR LITERATURE SEARCH

A literature search was conducted on Ovid MEDLINE (R) from 1946 to June Week 1 2016 and Embase Classic + Embase from 1947 to 2016 Week 24 databases using keywords including "inflammation," "cancer," "radiation," "neoplasm," "pain," "biological marker," "allele," and "genomics." The search generated a total of 5,480 results that were independently screened by two authors (MC, AF) first by title, then abstract, and subsequently full text using eligibility and ineligibility criteria. Figure 1 provides an overview of the screening process. If a disagreement arose on whether an article should be included, discussion ensued until a consensus could be reached. If a consensus was not reached a third author (LR) was consulted.

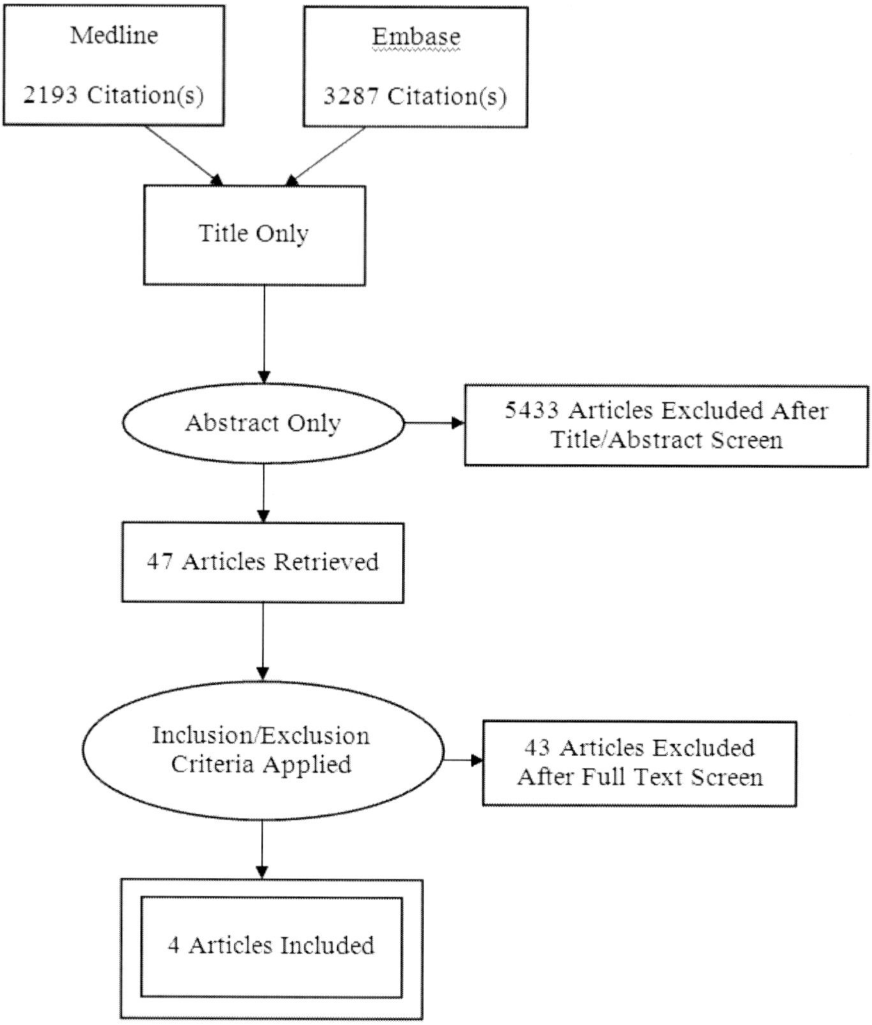

Figure 1. Summary of the literature screening process.

Screening criteria

Eligibility and ineligibility criteria were generated for the present review by four authors (MC, AF, LR, RM) to select articles for inclusion. Articles were included if they discussed biological or genetic biomarkers of radiation response, relating to radiation dose, fractionation, or patient outcome. Articles were excluded if they discussed outcomes of chemoradiotherapy exclusively or did not specify the biomarker examined. Articles were also excluded if they only mentioned side effects of radiation response and not patient outcome. Only original research articles were included in this review; case reports, case series and review articles were not included. The reference section of each review article identified was screened for potentially eligible studies.

Data extraction

Data extraction was completed independently by two authors (MC, AF) and subsequently compared to ensure accuracy. Information of interest was extracted from each full text article including characteristics of the biomarker examined, method of analyzing the biomarker, study cohort characteristics, main findings, and conclusions.

Table 1. Summary of the literature

Author (year)	Notes	Biomarker	Results	Citation
Barrett et al. (2014)	19 genes (52 SNPs) previously associated with radiation-induced toxicities were analyzed in 280 patients treated with curative radiation or chemoradiation for NSCLC to assess their correlation with overall survival	*IL1A*	*IL1A* SNP was significantly associated with overall survival (HR=0.58, 95%CI= 0.35-0.96; p=0.03) Number of minor allele variation copies correlated with improved overall survival but higher risk of pneumonitis	(3)
Schoenfeld et al. (2013)	3 SNPs within the *RNASEL* gene were correlated to the development of lethal prostate cancer or biochemical recurrence in 434 early-stage prostate cancer patients who received radiotherapy	*RNASEL* Variants: rs12757998 rs486907 rs627928	The rs12757998 variant was significantly associated with decreased risk of the composite endpoint (HR=0.65, 95%CI=0.45-0.94; p=0.02) and also biochemical recurrence alone (HR=0.60, 95%CI=0.40–0.89; p=0.01)	(2)

Table 1. (Continued)

Author (year)	Notes	Biomarker	Results	Cita-tion
Bortolin et al. (2014)	Correlated plasma levels of circulating inflammatory cytokines with local recurrence (LR) and distant failure (DF) rates in 22 early-stage NSCLC patients treated with stereotactic body radiation therapy (SBRT)	Cytokines: IL1β IL1R1 IL2 IL4 IL5 IL6 IL10 IL12(p40) IL13 IL15 IL17 INFγ TNFα Chemokines: IL8 MIP1α MIP1β Growth factors: IL7 EGF FGF2 TGFα VEGF	LR was significantly associated with EGF, IL10, and IL1β levels at baseline ($p<0.05$) DF was significantly associated with EGF, IL4, IL15, IL12, IL1β, IL1R1, INFγ and VEGF levels at baseline ($p<0.05$) Baseline levels of IL1R1, IL12, IL10, IL17, INFγ, and FGF2 were increased in NSCLC patients compared to healthy controls	(4)
Yu et al. (2015)	Analysed 9 miRNA levels from blood samples of 24 esophageal squamous cell carcinoma (ESCC) patients prior to, during, and following radiotherapy; these were correlated with radiographic response to radiation as well as 3-year overall survival	miR21 miR16 miR126 miR223 miR375 miR22 miR148b miR185 miR221	miR16 levels following radiotherapy were significantly higher in patients with good treatment outcomes versus poor treatment outcomes (p=0.038, fold change=2.85) A two-fold or higher increase in miR16 levels was a significant prognostic factor for overall survival in multivariate analysis (HR=0.263, 95%CI=0.093-0.745; p=0 .012)	(6)

WHAT WE FOUND

A total of four articles were included in the present selected review. Table 1 provides a summary of each of the studies.

Schoenfeld et al. analyzed 3 different single nucleotide polymorphisms (SNPs) (rs486907, rs627928, and rs12757998) in the *Ribonuclease L gene* (*RNASEL*) in 434 early-stage prostate cancer patients who underwent radiotherapy (2). The primary outcome of the study was to assess the correlation between these SNPs and a composite endpoint (development of lethal prostate cancer or biochemical recurrence) (2). Multivariate analysis showed that the rs12757998 variant allele and decreased risk of the composite endpoint were significantly associated (HR=0.65,95%CI=0.45-0.94; p=0.02) (2). This association was not significant when only the development of lethal prostate cancer was considered. However, it was significantly associated when only biochemical recurrence was considered (HR=0.60, 95%CI=0.40-0.89; p=0.01) (2).

Barrett et al. analyzed 19 genes with 52 SNPs in 280 patients treated with curative radiation or chemoradiation for non-small cell lung cancer (NSCLC) between 2005 and 2012 (3). The primary outcome of the study was to assess whether certain SNPs previously associated with radiation-induced toxicities could also be used for assessing differences in overall survival (3). They reported that an *IL1A* SNP was significantly associated with overall survival (HR=0.58, 95%CI=0.35-0.96, p=0.03) (3). This SNP has previously been associated with a higher risk of radiation pneumonitis (3). Additionally, their analysis suggested that patients with more copies of the minor allelic variation had a higher 3-year survival rate (72%, 82%, 90% for minor allele copy number 0, 1, and 2, respectively), but also suffered from higher rates of pneumonitis (15%, 22%, 48%) (3).

The study by Bortolin et al. aimed to correlate plasma levels of circulating inflammatory cytokines with local recurrence (LR) and distant failure (DF) rates in 22 early-stage NSCLC patients treated with stereotactic body radiation therapy (SBRT) (50-52 Gy in 5-8 fractions) (4). Blood samples were collected on the first and last day of radiotherapy, as well as on the 45th day following the end of SBRT (4). Their results showed that all NSCLC patients had increased baseline levels ($p<0.05$) of various different inflammatory cytokines (IL1R1, IL12, IL10, IL17, INFγ, and FGF2) at the start of treatment when compared to healthy controls (4). The study also found that IL1β, IL1R1, IL4, IL7, IL12, and FGF2 levels on the last day of treatment for patients who experienced LR or DF were significantly reduced from baseline ($p<0.05$), but returned to pre-treatment levels 45 days after the end of treatment (4). On univariate analysis for plasma cytokine levels at baseline, LR was significantly associated with EGF, IL10, and IL1B, while DF was significantly associated with EGF, IL4, IL15, IL12, IL1B, IL1R1, INFγ and VEGF ($p<0.05$) (4).

In order to determine the association between miRNA levels and radiation treatment outcomes, Yu et al. analyzed 9 different plasma miRNA levels in 24 patients with esophageal squamous cell carcinoma (ESCC) each of whom received approximately 32 sessions of radiotherapy (6). miRNA levels were measured from blood samples collected prior to radiotherapy, one week after the start of radiotherapy, and one day after the last session of radiotherapy (6). The treatment outcomes analyzed included tumour radiographic response based on CT scans and also 3-year overall survival (6). This study found that miR-16 levels following radiotherapy were significantly higher in patients with good treatment outcomes compared to those with poor treatment outcomes (p=0.038, fold change=2.85) (6). Additionally, a 2-fold or higher increase in miR-16 levels was a significant prognostic factor for overall survival in multivariate analysis (HR=0.263, 95%CI=0.093-0.745, p=0 .012) (6).

Discussion

As technology continues to improve, cancer researchers have the ever-growing ability to develop treatments tailored to individual patient and tumour characteristics. The field of cancer genetics is now expanding into looking at genetic variations and genetic biomarkers which may be implicated in predicting radiotherapy outcomes.

An interesting development in radiation therapy and cancer therapy is the use of microRNA (miRNA) (2,7). miRNA are highly conserved small RNA molecules consisting of 19-24 nucleotides that regulate gene expression (8–10). A single miRNA molecule can alter the gene expression of various target genes, and a single genetic target can be altered by many different miRNA molecules (8–10). Previous studies have shown that miRNA expression patterns vary between cancer cells and normal cells as well as between cancer types. Dysregulation of the miRNA expression is a distinctive feature of tumours and the surrounding microenvironment (11). In 2008, Yu et al, reported a distinct miRNA profile consisting of 5 different miRNA molecules that could be used to predict relapse and survival in patients with NSCLC (12). Their findings suggested that patients with higher-risk miRNA profile scores were associated with poorer overall and disease-free survival, compared to those with low-risk scores (12). A study conducted in 2015 by Yu et al. showed that ESCC patients with higher post-treatment miR-16 levels were associated with good treatment outcomes ($p=0.038$) and that a 2-fold or higher increase in miR-16 levels over the course of radiotherapy was associated with improved overall survival ($p=0.012$) (6). However, the mechanism by which these miRNA profiles affect patient outcomes is still not completely understood.

In addition to regulating gene expression, miRNAs can also act as transcription factors, and modulators of protein activity (9–11). Therefore, miRNA molecules may also alter the sensitivity of cells receiving radiation therapy. Using in vitro cell transfection assays, Lee et al. examined the radiation response of cancer cells with artificially increased levels of miRNA7 (miR7). They found that high miR7 levels were associated with down-regulation of many signaling molecules, including EGFR, Akt, ERK, and STAT3 (13). One of the effects of this down-regulation of EGFR signaling is the delayed response of DNA repair in irradiated cells (13). This study indicates that miRNAs may change a cell's sensitivity to radiation through altering intracellular signaling. The down-regulation of these signaling pathways may have implications on cellular activity and metabolism, and therefore be potential therapeutic targets or future areas of study. Other in vitro studies have shown that ectopic up-regulation of different miRNAs can increase the sensitivity of cancer cells to radiotherapy (14,15). These studies suggest that miRNA profiling, a non-invasive and cost-effective prognostic tool, may be able to predict whether or not radiation therapy will be effective in patients (11).

Most patients receiving radiation therapy will experience pain relief, but complete remission is confined to only about 30% of patients (16). Patients with the same diagnosis and treatment may have varying responses to radiation and experiences of pain, suggesting that treatment outcomes may be related to genetic variations within these patients. Our literature review highlighted 4 articles presenting biomarkers associated with radiation response. A common theme amongst these studies is that radiation causes an increase in inflammatory cytokines. Inflammation is a hallmark of radiation as it induces oxidative stress, DNA damage, and further expression of inflammatory genes (4,17,18). However, studies have reported contradictory outcomes where elevated levels of inflammatory markers are associated with both improved survival and decreased pain as well as poor survival and adverse side effects (4,19). These findings suggest that the type of cancer, location of tumours, genetic variants and potential mutations may be modulating inflammatory effects and outcomes of radiation therapy. Further genetic analysis should be done on various inflammatory proteins implicated in cancer for potential genetic variants that may be modulating the efficacy of patient outcomes (overall survival, disease-free survival, local recurrence).

Plasma biomarkers may potentially be related to variability in radiation response. Studies conducted by Bazan et al. (20) and Leung et al. (21) suggested that chemoradiation increases levels of circulating leukocytes, neutrophils in particular (20,21). Increases in inflammatory blood biomarkers such as absolute-neutrophil count (ANC), white blood cell count (WBC), and lymphocyte-to-neutrophil ratio (LNR) were prognostic of worse overall survival for patients receiving concurrent chemoradiation therapy (20,21). While there is a correlation between increased levels of inflammatory biomarkers and worse overall survival in patients receiving therapy, the mechanism by which this occurs is uncertain. Whether this relationship still exists for patients receiving only radiotherapy remains to be determined and should be further evaluated.

The articles cited in this review demonstrate potential associations between genetic biomarkers and radiotherapy outcomes as well as miRNA profiles and radiotherapy outcomes. However, the mechanism by which these biomarkers may be able to predict outcomes remains unclear. Additionally, differences in response to radiation may also be attributed to tumour location, disease stage, and comorbidities, yet not all of the included studies may have accounted for these factors. Due to a severe lack of literature surrounding this topic, further research is required to understand the relationship between these biomarkers, inflammation and radiation response. Future studies should seek to evaluate common pro-inflammatory genes and cytokines as well as miRNA profiles in order to build a better idea of their ability to predict treatment response. Medical professionals should continue to assess multiple biomarkers and use multidisciplinary approaches to determine the efficacy of radiation therapy. Ultimately, the goal of biomarker research remains to allow physicians to provide more effective patient-centered care.

ACKNOWLEDGMENT

We thank the generous support of Joey and Mary Furfari Cancer Research Fund.

References

[1] Delaney G, Jacob S, Featherstone C, Barton M. The role of radiotherapy in cancer treatment: Estimating optimal utilization from a review of evidence-based clinical guidelines. Cancer 2005;104(6):1129-37.
[2] Schoenfeld JD, Margalit DN, Kasperzyk JL, Shui IM, Nguyen PL, Kantoff PW, et al. A single nucleotide polymorphism in inflammatory gene RNASEL predicts outcome after radiation therapy for localized prostate cancer. Clin Cancer Res 2013;19(6):1612–9.
[3] Barrett K, Xu J, Kim S, Boyd K, Cheng D, Chen Z, et al. Genetic polymorphisms associated with toxicity are associated with overall survival following curative radiation for non-small cell lung cancer. Int J Radiat Oncol 2014;90(1):S811–2.
[4] Bortolin, Trovò M, Furlan C, Bidoli E, Basaglia G, Avanzo M, et al. Multiplexed plasma cytokine chemokine and growth factor profiling in early-stage non-small cell lung cancer patients undergoing stereotactic body radiation therapy. Int J Radiat Oncol 2014;90(1):S811–2.
[5] Grover SA, Coupai L, Zowall H, Rajan R, Trachtenberg J, Elhilali M, et al. The economic burden of prostate cancer in Canada: Forecasts from the Montreal Prostate Cancer Model. CMAJ 2000;162(7):987–92.
[6] Yu Q, Li B, Li P, Shi Z, Vaughn A, Zhu L, et al. Plasma microRNAs to predict the response of radiotherapy in esophageal squamous cell carcinoma patients. Am J Transl Res 2015;7(10):2060–71.
[7] Xu T, Liao Z, O'Reilly MS, Levy LB, Welsh JW, Wang LE, et al. Serum inflammatory miRNAs predict radiation esophagitis in patients receiving definitive radiochemotherapy for non-small cell lung cancer. Radiother Oncol 2014;113(3):379–84.
[8] Yanaihara N, Caplen N, Bowman E, Seike M, Kumamoto K, Yi M, et al. Unique microRNA molecular profiles in lung cancer diagnosis and prognosis. Cancer Cell 2006;9(3):189–98.
[9] Rothschild SI. MicroRNA therapies in cancer. Mol Cell Ther 2014;2(1):7.
[10] Reddy KB. MicroRNA (miRNA) in cancer. Cancer Cell Int. 2015;15(1):38.
[11] Price C, Chen J. MicroRNAs in cancer biology and therapy: Current status and perspectives. Genes Dis 2014;1(1):53–63.
[12] Yu SL, Chen HY, Chang GC, Chen CY, Chen HW, Singh S, et al. MicroRNA signature predicts survival and relapse in lung cancer. Cancer Cell 2008;13(1):48–57.
[13] Lee KM, Choi EJ, Kim IA. MicroRNA-7 increases radiosensitivity of human cancer cells with activated EGFR-associated signaling. Radiother Oncol 2011;101(1):171–6.
[14] Josson S, Sung S-Y, Lao K, Chung LWK, Johnstone PAS. Radiation modulation of microRNA in prostate cancer cell lines. Prostate 2008;68(15):1599–606.
[15] Weidhaas JB, Babar I, Nallur SM, Trang P, Roush S, Boehm M, et al. MicroRNAs as potential agents to alter resistance to cytotoxic anticancer therapy. Cancer Res 2007;67(23):11111–6.
[16] Koswig S, Budach V. Remineralization and pain relief in bone metastases after after different radiotherapy fractions (10 times 3 Gy vs. 1 time 8 Gy). A prospective study. Strahlenther Onkol 1999;175(10):500–8.

[17] Rodriguez-Gil JL, Takita C, Wright J, Reis IM, Zhao W, Brian EL, et al. Inflammatory biomarker c-reactive protein and radiotherapy-induced early adverse skin reactions in breast cancer patients. Cancer Epidemiol Biomarkers Prev 2014;23(9):1873–83.
[18] Yoritsune E, Furuse M, Kuwabara H, Miyata T, Nonoguchi N, Kawabata S, et al. Inflammation as well as angiogenesis may participate in the pathophysiology of brain radiation necrosis. J Radiat Res 2014;55(4):803–11.
[19] Meyer MS, Penney KL, Stark JR, Schumacher FR, Sesso HD, Loda M, et al. Genetic variation in RNASEL associated with prostate cancer risk and progression. Carcinogenesis 2010;31(9):1597–603.
[20] Bazan JG, Allan E, Washington I, Majithia L, Grecula JC, Haglund KE, et al. Increased pretreatment neutrophil-to-lymphocyte ratios and platelet-to-lymphocyte ratios are independent prognostic markers for worse overall survival in patients receiving concurrent chemoradiation for lung cancer. Int J Radiat Oncol 2015;93(3):E440.
[21] Leung EW, Pintilie M, Yap M, Cuartero J, MacKay H, Clarke B, et al. Neutrophils Modulate Vascular Function in Locally-Advanced Cervical Cancer and Impair Response to Radiation Therapy. Int J Radiat Oncol Biol Phys 2013;87(2):S31–2.

In: Oncology
Editors: A. Furfari et al.
ISBN: 978-1-53610-608-4

Chapter 7

BIOMARKERS OF RADIATION RELATED TOXICITIES

Anthony Furfari[1], MD(C), Stephanie Chan[1], BSc(C), Stephanie Cheon[1], MD(C), Pearl Zaki[1], BSc(C), Bo Angela Wan[1], MPhil, Leigha Rowbottom[1], MD(C), Rachel McDonald[1], MD(C), Azar Azad[2,3,4], PhD, Ronald Chow[1], BMSc(C), Carlo DeAngelis[1], PharmD, Michael Lam[1], BMSc(C), Henry Lam[1], MLS, Edward Chow[1], MBBS and George S Charames[3,4,5,*], PhD

[1]Odette Cancer Centre, Sunnybrook Health Sciences Centre, University of Toronto, Toronto, Ontario, Canada

[2]Mount Sinai Services Inc., Toronto, Ontario, Canada

[3]Pathology and Laboratory Medicine, Mount Sinai Hospital, Toronto, Ontario, Canada

[4]Laboratory Medicine and Pathobiology, University of Toronto, Toronto, Ontario, Canada

[5]Lunenfeld-Tanenbaum Research Institute, Sinai Health System, Toronto, Ontario, Canada

[*] Correspondence: George S Charames PhD, FACMG, Pathology and Laboratory Medicine, Mount Sinai Hospital, Suite 6-500, 600 University Ave, Toronto, Ontario, M5G 1X5, Canada. E-mail: George.Charames@sinaihealthsystem.ca.

Radiation therapy (RT) is a commonly used modality to treat cancer patients with either curative or palliative intent. However, RT may induce toxicities that range from being negligible to severe. Sometimes, radiation-induced toxicities may be dose-limiting, preventing patients from completing treatment and receiving the full benefit of RT. Genetic and protein biomarkers could potentially allow physicians to predict the risk of each patient developing a toxicity, thus improving patient-centred care and providing patients with more information to help them make treatment-related decisions. In this chapter we summarize the findings of all human studies analysing the potential of different biomarkers in predicting the risk of toxicities in cancer patients receiving radiotherapy. In the present literature review, a search was conducted on Medline and Embase databases using keywords including "inflammation," "cancer," "radiation," "neoplasm," "cancer," "pain," "biological marker," "allele," and "genomics". The search generated 5,480 and seven were included in the final review. Several studies were able to find genetic variants significantly associated with increased or decreased risk of radiation-induced toxicities, and even generated risk scores that improved current models to more accurately predict this risk. Future studies should analyse more genes in larger sample sizes while considering factors including ethnicity, patient characteristics, and dosimetric factors.

INTRODUCTION

Radiation therapy (RT) is a non-invasive method of combatting cancer in the body and has proven itself to be an invaluable tool for treating cancer patients whether with palliative or curative intent. It has been estimated that 52% of all cancer patients require RT at some point (1). However, RT can also cause toxicities such as adverse skin reactions in patients receiving adjuvant radiation for breast cancer or esophagitis and pneumonitis in lung cancer patients. Such adverse events occur as a result of the radiation beam damaging nearby normal tissues rather than malignant cells. These toxicities can severely impact patients' quality of life (QOL), subsequently causing the interruption or cessation of treatment. Thus, radiation-induced toxicities should be avoided whenever possible and finding novel methodologies to improve the therapeutic index is a focus of clinical research.

It is well known that not all patients will experience the same degree of toxicities, even if they are identical in terms of prognostic clinical variables such as diagnosis, age, sex, and performance status. This may be suggestive of a genetic component in radiation-induced toxicities. Genetic and protein biomarkers could potentially be a novel and more accurate method of predicting individual risk to certain toxicities. Implementation of this method in the clinical setting could improve personalized patient care, allowing both physicians and patients to make more informed decisions and have more accurate expectations regarding treatment. The aim of the present review is to summarize the current literature surrounding protein and genetic biomarkers in predicting susceptibility to radiation-induced toxicities in cancer patients.

OUR SEARCH

A comprehensive literature search was conducted in Ovid MEDLINE (R) spanning 1946 to June Week 1 2016 and Embase Classic + Embase spanning 1947 to 2016 Week 24. The keywords "inflammation," "cancer," "radiation," "neoplasm," "cancer," "pain," "biological marker," "allele," and "genomics" were used. Results were limited to studies conducted in humans and English language. After removing duplicated articles, the literature search was screened by two authors (AF, SC) independently based on the title, abstract, and then full text according to the pre-defined eligibility and ineligibility criteria. If discrepancies arose between the two authors, discussion ensued until a decision could be made. Studies were included in the literature review if they examined biomarkers of radiation-induced toxicities in cancer patients receiving radiotherapy. Case reports, literature reviews, editorials, or studies not in the English language were excluded. Studies regarding side effects of radiopharmaceuticals or chemoradiation therapy only were excluded. After all studies were chosen, data including the purpose of the studies, the sampling and analysis methods used, the study population, the findings and conclusions, and the biomarkers analysed were extracted.

OUR FINDINGS

The literature search returned 5,480 articles, with 3,287 from MEDLINE, and 2,193 from EMBASE, of which 7 were identified for this review (3–9). These studies are summarized in Table 1.

Table 1. Summary of included studies

First Author (Year)	Aim	Biomarker	Result	Ref.
Rodriguez-Gil et al. (2014)	Evaluated C-Reactive protein (CRP) levels from blood samples of 159 breast cancer patients receiving RT following surgery as a biomarker of radiation-induced early adverse skin reactions (EASR)	C-reactive Protein (CRP)	No significant association between EASRs and pre-RT CRP levels Pre-RT CRP levels combined with obesity was significantly correlated to grade 2+ EASRs 6 weeks post-RT (OR=3.03, 95% CI=1.28–7.17, p=0.01) Grade 2/3 EASRs 6 weeks post-RT significantly associated with higher post-RT CRP levels combined with obesity (OR=2.70, 95% CI=1.18–6.19, p=0.02) and change in CRP levels (OR=2.51, 95% CI=1.06–5.95, p=0.04)	(3)
Ho et al. (2007)	Examined association between variants in *ATM* and radiation-induced skin and subcutaneous tissue reactions in 131 breast cancer patients receiving adjuvant radiotherapy following breast surgery	*ATM*	Patients with any *ATM* variant (n=51) were at a significantly higher risk of developing a grade 2+ late toxicity compared to those with no sequence variation (n=80) (OR=2.4, 95% CI=1.1-5.2) Patients heterozygous for the *5557G→A* polymorphism (n=15) exhibited significantly higher rates of grade 2+ late toxicity compared to those without it (n=116) (OR=3.1, 95% CI=1.1-9.4)	(6)
Pu et al. (2014)	Examined correlation between radiation-induced pneumonitis and esophagitis in 201 Caucasian non-small cell lung cancer (NSCLC) patients receiving radiotherapy or chemoradiation. 11,930 SNPs were analyzed	Esophagitis: *OSMR* - rs1239344 *TNFSF7* - rs7259857 *PRKCE* - rs940052 *FGF14* - rs4772468 *TAP1* - rs3819721 *CD4* - rs2707212 *LILRP2* - rs270771	10 SNPs significantly associated with increased risk of esophagitis, most significant was rs1239344 in *OMSR* which was associated with a 2-fold increased risk in the discovery (OR=2.45, 95% CI=1.14-5.26, p=0.021) and validation populations (OR=4.15, 95% CI=1.68–10.28, p=0.002) 9 SNPs significantly associated with higher risk of pneumonitis, most significant was rs10711 in *CDK1* which was associated with a 2-fold increased risk in the discovery (OR=2.67, 95%	(4)

	and significant SNPs were validated with an additional 220 NSCLC patients	*IL15RA* - rs1998521 *TANK* - rs7309 *AGER* - rs204993 Pneumonitis: *CDC2* - rs10711, rs1871445 *DDX58* - rs11795343, rs7865082 *FGF5* - rs3733336 *ETS2* - rs2298560 *LIMS1* - rs12469016 *GHR* - rs4292454 *TFEB* - rs13202921	CI=1.26–5.63, p=0.010) and validation phases (OR=2.33, 95% CI=1.21–4.48, p=0.011) Polygenic risk scores were generated based on the 19 significantly associated SNPs which was found to significantly improve the discriminative ability of the clinical variable-based model in predicting risk of radiation-induced toxicities. 45 SNPs in the *PRKCE*, *DDX58*, and *TNFSF7* genes were significantly associated with radiation response ($p=1.21\times10^{-27}$)	
Hildebrandt et al. (2010)	Examined 59 pro-inflammatory and anti-inflammatory related SNPs and their association with radiation-induced esophagitis or pneumonitis in Caucasian 173 NSCLC patients receiving radiotherapy or chemoradiation	*IL6* - rs1800795 *IL16* – rs11556218 *TNF* – rs1799724 *PTGS2* - rs20417, rs5275, rs689470 *IL4R* – rs1801275 *IL10* – rs1800872 *IL10RA* - rs3135932 *IL1A* – rs1800587, rs17561 *IL8* – rs4073 *TNFRSF1B* - rs1061622 *MIF* - rs755622 *NOS3* - rs1799983 *IL4* - rs2243250, rs2070874 *IL13* - rs20541, rs180925 *NFKBIA* - rs8904	1.58-3.38-fold increased risk of esophagitis significantly associated with pro-inflammatory genes (*IL6, IL16, TN, PTGS2*) 1.49-4.12-fold increased risk of esophagitis significantly associated with anti-inflammatory genes (*IL4R, IL10, IL10RA*) 2.12-4.96-fold increased risk of pneumonitis significantly associated with some pro-inflammatory genes (*IL1A, IL8, TNF, TNFRSF1B, MIF*) except for *NOS3* which was associated with a 45% decreased risk of pneumonitis (p=0.037) 2.02-3.23-fold increased risk of pneumonitis significantly associated with anti-inflammatory genes (*IL4, NFKBIA, IL13*) Effect of SNPs on risk of pneumonitis or esophagitis was found to be cumulative ($p<0.001$)	(5)

Table 1 (Continued)

First Author (Year)	Aim	Biomarker	Result	Ref.
Guerra et al. (2012)	Examined 3 SNPs in 1 gene and correlated to risk of developing radiation-induced esophageal toxicities in 97 NSCLC patients who received radiotherapy or chemoradiation and validated this in an additional 101 NSCLC patients	*TGFB1* - rs1800469, rs1800471, and rs1982073	Patients with the *CT/TT* genotype in the rs1800469 variant were significantly associated with increased risk of esophageal toxicities of grade 3 or higher in multivariate analysis in both the test set (HR=2.47, 95% CI=1.02–6.02, p=0.045) and the validation set (HR=2.53, 95% CI=1.14–5.63, p=0.023) Patients with 3 unfavourable genotypes were at higher risk of grade 3 or higher esophageal toxicities than patients with only 1 (test set, p = 0.013; validation set, p = 0.004)	(7)
Yuan et al. (2009)	Investigated the association between SNPs in transforming growth factor 1 (*TGFB1*) gene and the risk of pneumonitis in NSCLC patients	*TGFB1* - rs1800469, rs1800471, and rs1982073	the *CT/CC* rs1982073 genotype was significantly associated with decreased risk of pneumonitis on multivariate analysis (grade ≥2, HR=0.489, 95% CI=0.227-0.861, p=0.013; grade ≥3, HR=0.390, 95% CI=0.197-0.774, p=0.007) compared to the TT genotype	(9)
Mohammed et al. (2012)	Analyzed erythrocyte sedimentation rate (ESR) and C-Reactive protein (CRP) measurements before, during, and following RT in 62 head and neck cancer patients treated with RT or concurrent chemoradiation	C-Reactive protein (CRP) erythrocyte sedimentation rate (ESR)	ESR significantly increased at week 3 after the start of RT until end of RT (p=0.01) CRP levels significantly increased from week 6 (p=0.01) until week 8 (p=0.03)	(8)

Toxicities in Breast Cancer Patients

The study by Rodriguez-Gil et al. aimed to establish the association between C-Reactive protein (CRP) and radiation-induced early adverse skin reactions (EASR) in 159 breast cancer patients receiving RT following surgery (3). Patients received regular doses of 45-50.4 Gy in 1.8 or 2.0 Gy per fraction to the whole breast and CRP levels were obtained from blood samples collected immediately prior to and following the completion of RT. EASRs were assessed 3 and 6 weeks following the completion of RT. After adjusting for age, race, ethnicity, and tumour stage, the study observed no significant associations between EASRs and pre-RT CRP levels. However, grade 2 or higher EASRs 6 weeks following RT were significantly associated to pre-RT CRP levels when combined with obesity (odds ratio (OR) = 3.03, 95% confidence interval (CI) = 1.28–7.17, p = 0.01). Additionally, grade 2 or 3 EASRs 6 weeks following RT were significantly associated with higher post-RT CRP levels combined with obesity (OR = 2.70, 95% CI = 1.18–6.19, p = 0.02), change in CRP levels (OR = 2.51, 95% CI = 1.06–5.95, p = 0.04), and obesity (OR = 2.08, 95% CI = 1.03–4.21, p = 0.04) (3).

Ho et al. examined the association between genetic variants in the cell cycle regulation gene, *ataxia telangiectasia mutated* (*ATM*), and radiation-induced skin or subcutaneous tissue reactions using blood samples from 131 breast cancer patients receiving adjuvant RT following breast surgery (6). The authors found that patients with any *ATM* variant (n = 51) were at a significantly higher risk of developing a grade 2 or higher late radiation toxicity compared to those with no sequence variation in this gene (n = 80) (OR = 2.4, 95% CI = 1.1-5.2). Additionally, patients heterozygous for the benign *ATM* c.5557G > A (p.Asp1853Asn) missense variant (n = 15) exhibited significantly higher rates of grade 2 or higher late radiation toxicity compared to those with only wild type alleles (n = 116) (OR = 3.1, 95% CI = 1.1-9.4) (6).

Pneumonitis and esophagitis in NSCLC patients

The study by Pu et al. examined the correlation between radiation-induced toxicities, specifically pneumonitis and esophagitis, in 201 Caucasian non-small cell lung cancer (NSCLC) patients receiving definitive radiotherapy or concurrent chemoradiation therapy (4). This was done by analysing a total of 11,930 single nucleotide polymorphisms (SNPs), of which 19 remained significant after validation with an additional 220 NSCLC patients. The most significant of the 10 SNPs significantly associated with esophagitis was rs1239344 located in the *OMSR* gene. A two-fold increased risk of esophagitis was associated with this SNP in both the discovery (OR = 2.45, 95% CI = 1.14-5.26, p = 0.021) and validation populations (OR = 4.15, 95% CI = 1.68–10.28, p = 0.002), and was associated with a 3-fold increase risk of esophagitis in the combined meta-analysis (p = 1.78×10^{-4}, 95% CI = 1.70-5.57). When pneumonitis was analysed, significant associations were found with 9 SNPs. Significant associations between higher risk of pneumonitis and rs10711 of *CDK1* in the discovery (OR = 2.67, 95% CI = 1.26–5.63, p = 0.010) and validation phases (OR = 2.33, 95% CI = 1.21–4.48, p = 0.011). This remained significant in the meta-analysis (OR = 2.47, 95% CI = 1.51–4.04, p = 3.08×10^{-4}). The authors also generated a polygenic risk score based on risk genotypes that was able to significantly improve the ability of the baseline clinical variable-based risk model to predict which patients were at higher risk of radiation-induced toxicities. Additionally, 45 SNPs in the *PRKCE, DDX58,* and *TNFSF7* genes were significantly associated with radiation response (p = 1.21×10^{-27}) (4).

Hildebrandt et al. also examined the association between pneumonitis or esophagitis and SNPs of inflammation-related genes in Caucasian NSCLC patients treated with radiotherapy or chemoradiation therapy (5). Of the 59 SNPs from 37 different genes analysed in 173 NSCLC patients, 6 SNPs of pro-inflammatory genes (*IL6, IL16, TNF, PTGS2*) and variants of 3 anti-inflammatory genes (*IL4R, IL10, IL10RA*) were significantly associated with increased esophagitis risk. Variants in 6 pro-inflammatory genes (*IL1A, IL8, TNF, TNFRSF1B, MIF, NOS3*) as well as 3 anti-inflammatory genes (*IL4, IL13, NFKBIA*) were significantly associated with increased risk of pneumonitis. One pro-inflammatory gene, *NOS3,* was significantly associated with a decreased risk of pneumonitis (HR = 0.55, 95% CI = 0.31–0.96). Combined analysis revealed that patients with 4 and 5 or greater unfavourable genotypes had a 3.71-fold increased risk and 8.85-fold increased risk of esophagitis, respectively, compared to patients with 3 or less unfavourable genotypes. The increased risk of pneumonitis for patients with 3 unfavourable genotypes was 13.30-fold compared to patients with 2 or less unfavourable genotypes (p = 0.013). For patients with 4 or more unfavourable genotypes, the risk of pneumonitis was dramatically increased ($p < 0.0001$) (5).

The study by Guerra et al. examined the association between 3 SNPs (rs1800469, rs1800471, and rs1982073) in *the transforming growth factor beta 1* gene (*TGFB1*) and esophageal toxicities in 97 NSCLC patients treated with RT or chemoradiation therapy treated between 2003 and 2006 (7). They also analyzed this in a validation set of 101 NSCLC patients treated between 1998 and 2002. They found that patients with the CT/TT genotype in the rs1800469 variant were significantly associated with increased risk of esophageal toxicities of grade 3 or higher in multivariate analysis in both the test set (HR = 2.47, 95% CI = 1.02–6.02, p = 0.045) and the validation set (HR = 2.53, 95% CI = 1.14–5.63, p = 0.023). Additionally, they found that patients with 3 unfavourable genotypes were at higher risk of grade 3 or higher esophageal toxicities than patients with only one unfavourable genotype (test set p = 0.013, validation set p = 0.004) (7).

Yuan et al. analyzed 3 SNPs (rs1800469, rs1800471, rs1982073) of *TGFB1* in 164 genomic samples from NSCLC patients treated with definitive radiation or concurrent chemoradiation (9). The primary outcome was to assess the association between SNPs in *TGFB1* and the risk of developing grade ≥2 or grade ≥3 pneumonitis in this population. The authors found that the CT/CC rs1982073 genotype was significantly associated with decreased risk of pneumonitis on multivariate analysis (grade ≥2 HR = 0.489, 95% CI = 0.227-0.861, p = 0.013; grade≥3 HR = 0.390, 95% CI = 0.197-0.774, p = 0.007) compared to the TT genotype (9).

Mucositis in head and neck cancer patients

The study by Mohammed et al. included 62 head and neck cancer patients treated with either RT or concurrent chemoradiation with a radiation dose of at least 50Gy between 2002 and 2005 (8). The primary outcome of the study was to correlate erythrocyte sedimentation rate (ESR) and CRP measurements at time of enrolment, once a week during RT, and 1 month following completion of RT to mucositis, a dose-limiting radiation-induced toxicity in head and neck cancer patients. They found that at week 3 after the start of radiation, ESR significantly increased (p = 0.01) until the end of RT. Additionally, CRP levels were significantly higher than baseline starting from week 6 post-RT (p = 0.0002) until week 8 (p = 0.03). These rises in CRP and ESR over treatment time may correspond to the onset of toxicities, however this direct association was not confirmed in the study (8).

Discussion

In the present literature review, seven studies analysing the efficacy of genetic and protein biomarkers in predicting toxicities in cancer patients receiving curative or palliative RT were summarized. When patients are treated with radiation, inflammation of the local tissues occurs as a natural response to tissue damage, ultimately causing toxicities. This inflammatory response is regulated by the production of anti-inflammatory and pro-inflammatory cytokines (10,11). Although higher doses of radiation have been linked to improved outcomes as more malignant cells are damaged (12), toxicities may be dose-limiting factors preventing patients from receiving the full treatment and subsequently decreasing the efficacy of RT. Currently, no clinical models are able to reliably predict patients' risk of developing these toxicities and, as made evident by the papers included in the present review, biomarkers may have the potential to do so.

As individual genetic variations in inflammatory cytokine levels may be associated with patient-specific risk of developing toxicities (2), analysing circulating inflammatory cytokines in cancer patients receiving treatment as biomarkers of radiation-induced toxicities is a research topic of interest. For example, the study included in this review by Rodriguez-Gil et al. examined if CRP levels prior to and following RT in breast cancer patients were able to predict the prevalence of EASRs, and found significant associations between the two when combined with obesity as a factor (3). CRP levels have also been associated with other toxicities including mucositis in head and neck cancer patients (8), and was prognostic in breast cancer patients (13). However, very few studies analysing CRP levels and other serum protein levels have been able to produce reliable results. Additionally, using changes in cytokine levels before and after RT to predict risk of toxicities has little clinical value as the patient would need to receive RT before physicians could inform patients of their risk profile. Thus, some researchers have altered their focus towards genetic biomarkers. Specifically, SNPs are an attractive area of research due to their heritable nature as well as their capacity to yield reproducible results.

Several studies have been conducted in NSCLC patients receiving RT to the lung, a procedure that puts them at risk of developing pneumonitis and esophagitis. Inflammation is a key cellular process in pneumonitis and esophagitis, thus many studies have focused on analysing the genes' corresponding SNPs implicated in inflammation. In the study by Pu et al, many of the SNPs that were significantly associated with toxicity also play key roles in inflammation pathways such as the mitogen-activated protein kinase, Janus kinase, nuclear factor κB, and other transcription pathways (4). These genes have even been associated with radiation response (14,15). In the included studies, most SNPs found to be significant in predicting toxicities were associated with increased risk, while only a few demonstrated decreased risk. One example is the study by Hildebrandt et al. in which the rs1799983 variant of *NOS3* was correlated with a 45% decreased risk of developing pneumonitis (5). This may be a result of reductions in nitric oxide production (16) due to the presence of a variant allozyme that reduces enzymatic activity in some patients (17), indicating decreasing inflammatory signaling.

Additionally, several studies included in this review have looked beyond the effects of individual SNPs on patients' risk of developing toxicities, and the focus has shifted towards looking at the cumulative effects of all SNPs through the development of a polygenic risk score (PRS) or a risk model that incorporates clinical variables as well as genetic biomarkers. Hildebrandt et al. found that the cumulative effect of the individual SNPs was still a significant predictor of toxicity risk (5). In the study by Pu et al, a PRS was generated based on the SNPs significantly associated with either increased pneumonitis or esophagitis in NSCLC patients (4). They found that their PRSs were able to reliably predict which patients were at high risk of developing toxicities when tested in both the discovery and validation phase (4). Risk scores or profiles can be used by physicians to easily provide patients with a comprehensive value that conveys relevant and individualized information regarding treatment and potential toxicities, thereby demonstrating significant clinical value. Once further research can determine which SNPs are significant predictors and their effects on patients' risk, risk scores can easily be generated and interpreted due to its cumulative nature. These risk scores may also be combined with clinical and dosimetric variables to make the prediction model even more accurate as previous models based on clinical variables only have had little success in accurately predicting risk (18–21).

A great potential exists for genetic biomarkers to improve current radiation-induced toxicity risk models, but further research must be done to make this a reality. Future studies should include larger sample sizes and analyse even more inflammation-related genes with both discovery and validation phases to improve precision and repeatability. In addition, researchers must also take into account race when conducting studies as ethnic genetic differences may play a role in radiation-induced toxicities (22,23). Clinical variables including age, tumour location, tumour size, radiation field size, total radiation dose, dose to organs at risk, and performance status should be incorporated in analyses. Finally, to elucidate the mechanisms by which genes and inflammatory proteins may affect toxicities in cancer patients, more biochemical research must be done.

ACKNOWLEDGMENTS

We thank the generous support of the Joey and Mary Furfari Cancer Research Fund.

REFERENCES

[1] Delaney G, Jacob S, Featherstone C, Barton M. The role of radiotherapy in cancer treatment: estimating optimal utilization from a review of evidence-based clinical guidelines. Cancer 2005;104(6):1129–37.

[2] Barnett GC, West CML, Dunning AM, Elliott RM, Coles CE, Pharoah PDP, et al. Normal tissue reactions to radiotherapy: towards tailoring treatment dose by genotype. Nat Rev Cancer 2009;9(2):134–42.

[3] Rodriguez-Gil JL, Takita C, Wright J, Reis IM, Zhao W, Lally BE, et al. Inflammatory biomarker C-reactive protein and radiotherapy-induced early adverse skin reactions in patients with breast cancer. Cancer Epidemiol Prev Biomarkers 2014;23(9):1873-83.

[4] Pu X, Wang L, Chang JY, Hildebrandt MAT, Ye Y, Lu C, et al. Inflammation-related genetic variants predict toxicity following definitive radiotherapy for lung cancer. Clin Pharmacol Ther 2014;96(5):609–15.

[5] Hildebrandt MAT, Komaki R, Liao Z, Gu J, Chang JY, Ye Y, et al. Genetic variants in inflammation-related genes are associated with radiation-induced toxicity following treatment for non-small cell lung cancer. PLoS One 2010;5(8):e12402.

[6] Ho AY, Fan G, Atencio DP, Green S, Formenti SC, Haffty BG, et al. Possession of ATM sequence variants as predictor for late normal tissue responses in breast cancer patients treated with radiotherapy. Int J Radiat Oncol Biol Phys 2007;69(3):677–84.

[7] Guerra JLL, Gomez D, Wei Q, Liu Z, Wang LE, Yuan X, et al. Association between single nucleotide polymorphisms of the transforming growth factor β1 gene and the risk of severe radiation esophagitis in patients with lung cancer. Radiother Oncol 2012;105(3):299–304.

[8] Mohammed FF, Poon I, Zhang L, Elliott L, Hodson ID, Sagar SM, et al. Acute-phase response reactants as objective biomarkers of radiation-induced mucositis in head and neck cancer. Head Neck 2012;34(7):985–93.

[9] Yuan X, Liao Z, Liu Z, Wang LE, Tucker SL, Mao L, et al. Single nucleotide polymorphism at rs1982073:T869C of the TGFbeta 1 gene is associated with the risk of radiation pneumonitis in patients with non-small-cell lung cancer treated with definitive radiotherapy. J Clin Oncol 2009;27(20):3370–8.

[10] Opal SM, DePalo VA. Anti-inflammatory cytokines. Chest 2000;117(4):1162–72.

[11] Dinarello CA. Proinflammatory cytokines. Chest 2000;118(2):503–8.

[12] Machtay M, Bae K, Movsas B, Paulus R, Gore EM, Komaki R, et al. Higher biologically effective dose of radiotherapy is associated with improved outcomes for locally advanced non-small cell lung carcinoma treated with chemoradiation: an analysis of the Radiation Therapy Oncology Group. Int J Radiat Oncol Biol Phys 2012;82(1):425–34.

[13] Allin KH, Nordestgaard BG. Elevated C-reactive protein in the diagnosis, prognosis, and cause of cancer. Crit Rev Clin Lab Sci 2011;48(4):155–70.

[14] Dent P, Yacoub A, Contessa J, Caron R, Amorino G, Valerie K, et al. Stress and radiation-induced activation of multiple intracellular signaling pathways. Radiat Res 2003;159(3):283–300.

[15] Valerie K, Yacoub A, Hagan MP, Curiel DT, Fisher PB, Grant S, et al. Radiation-induced cell signaling: inside-out and outside-in. Mol Cancer Ther 2007;6(3):789–801.

[16] Veldman BA, Spiering W, Doevendans PA, Vervoort G, Kroon AA, de Leeuw PW, et al. The Glu298Asp polymorphism of the NOS 3 gene as a determinant of the baseline production of nitric oxide. J Hypertens 2002;20(10):2023–7.

[17] Wang XL, Sim AS, Wang MX, Murrell GA, Trudinger B, Wang J. Genotype dependent and cigarette specific effects on endothelial nitric oxide synthase gene expression and enzyme activity. FEBS Lett 2000;471(1):45–50.

[18] De Ruyck K, Sabbe N, Oberije C, Vandecasteele K, Thas O, De Ruysscher D, et al. Development of a multicomponent prediction model for acute esophagitis in lung cancer patients receiving chemoradiotherapy. Int J Radiat Oncol Biol Phys 2011;81(2):537–44.

[19] Dehing-Oberije C, De Ruysscher D, van Baardwijk A, Yu S, Rao B, Lambin P. The importance of patient characteristics for the prediction of radiation-induced lung toxicity. Radiother Oncol 2009;91(3):421–6.

[20] Jenkins P, Watts J. An improved model for predicting radiation pneumonitis incorporating clinical and dosimetric variables. Int J Radiat Oncol Biol Phys 2011;80(4):1023–9.

[21] Zhu SC, Shen WB, Liu ZK, Li J, Su JW, Wang YX. Dosimetric and clinical predictors of radiation-induced lung toxicity in esophageal carcinoma. Tumori 2011;97(5):596–602.

[22] Brown MD, Feairheller DL, Thakkar S, Veerabhadrappa P, Park JY. Racial differences in tumor necrosis factor-α-induced endothelial microparticles and interleukin-6 production. Vasc Health Risk Manag 2011;7:541–50.

[23] Park NJ, Kang DH. Inflammatory cytokine levels and breast cancer risk factors: racial differences of healthy Caucasian and African American women. Oncol Nurs Forum 2013;40(5):490–500.

In: Oncology
Editors: A. Furfari et al.

ISBN: 978-1-53610-608-4

Chapter 8

COMMON INFLAMMATORY BIOMARKERS OF CANCER PROGNOSIS

Bo Angela Wan*[1]*, MPhil, Michael Lam*[1]*, BMSc(C), Anthony Furfari*[1]*, MD(C), Rachel McDonald*[1]*, MD(C), Leigha Rowbottom*[1]*, MD(C), Stephanie Chan*[1]*, BSc(C), Pearl Zaki*[1]*, BSc(C), Carlo DeAngelis*[1]*, PharmD, Ronald Chow*[1]*, BMSc(C), George S Charames*[2,3,4]*, PhD, Henry Lam*[1]*, MLS, Edward Chow*[1]*, MBBS and Azar Azad*[2,3,5,*]*, PhD

[1]Odette Cancer Centre, Sunnybrook Health Sciences Centre,
University of Toronto, Toronto, Ontario, Canada
[2]Pathology and Laboratory Medicine,
Mount Sinai Hospital, Toronto, Ontario, Canada
[3]Laboratory Medicine and Pathobiology,
University of Toronto, Toronto, Ontario, Canada
[4]Lunenfeld-Tanenbaum Research Institute,
Sinai Health System, Toronto, Ontario, Canada
[5]Mount Sinai Services Inc., Toronto, Ontario, Canada

[*] Correspondence: Azar Azad PhD, Pathology and Laboratory Medicine, Mount Sinai Hospital, 600 University Ave, Toronto, Ontario, Canada. E-mail: Azar.Azad@sinaihealthsystem.ca.

Biomarkers have the potential to act as important tools in predicting prognosis of cancer patients. Blood is one of the most readily available sources of biomarkers. In addition to being easy to obtain, levels of the protein and cell constituents in blood provide valuable information regarding the degree of inflammation. Inflammatory proteins, leukocyte levels, and platelet levels may be able to predict survival in many cancer types. In this chapter we summarize the currently available literature reporting on common inflammatory blood biomarkers, excluding cytokines and chemokines, and their association with prognosis in cancer patients. A literature search was conducted on Medline and Embase utilizing keywords such as 'neoplasm,' 'inflammation,' 'biomarker,' 'allele,' and 'genomics.' Articles that reported on the levels of commonly implemented prognostic biomarkers (C-reactive protein (CRP), albumin, white blood cell and platelet counts) and their derivative scores or ratios relating to prognosis were selected for inclusion. Information regarding the patient population, cancer type, interventions received, type of biomarker, and the impact on prognosis was extracted from each of the 23 studies included in this review. The most commonly reported biomarker was neutrophil-lymphocyte ratio, which was found to be significant across 12 studies spanning at least six different types of malignancies. Blood biomarkers associated with inflammation are a common prognostic tool due to the ease of blood sampling and its widespread significance across a range of cancers.

INTRODUCTION

Multiple studies investigating the mechanism of cancer development and the tumour microenvironment have established inflammation as one of the hallmarks of cancer (1). Such studies attribute several pathways by which inflammation contributes to cancer development. For instance, in normal tissue damage, immune cells facilitate repair by triggering the release of growth factors and proteins to stimulate angiogenesis. In the context of a tumour environment, such processes enhance the development of tumours by increasing access to nutrients which they can then consume (2). Moreover, the release of reactive oxygen species (ROS) by immune cells provides a mutagenic environment that facilitates genetic instability, therefore enabling the acquisition of mutations that allow cancer cells to rapidly divide without restriction, evade immune detection, and acquire metastatic potential (1).

Certain proteins act as measurable parameters of inflammation in the blood and have been established as significant predictors of prognosis in many cancer types. For instance, an elevation of the acute-phase protein C-reactive protein

(CRP) and a reduction of albumin (hypoalbuminemia) are indicative of an underlying systemic inflammatory response (3). The combined assessment of both CRP and albumin led to the development of the Glasgow Prognostic Score (GPS), which has been independently associated with prognosis in multiple cancer types in over 60 studies (3). The GPS assigns one point if CRP is greater than 10mg/L, one point if albumin is less than 35mg/L (hypoalbuminemia), and points for both high CRP and hypoalbuminemia; high GPS scores correlate to increased inflammation and worse prognosis. Since hypoalbuminemia without increased CRP was both rare and not found to be associated with survival by itself, in the modified GPS (mGPS), hypoalbuminemia without elevated CRP does not warrant a point.

As inflammation is an aspect of an immune response, it is associated with altered levels of white blood cells. Therefore, the level of circulating white blood cells and other associated parameters can be measured and used as indicators of inflammation. Such parameters include absolute counts of lymphocytes, such as the absolute neutrophil count (ANC), white blood cell count (WBC), and the absolute monocyte count (AMC). In addition, counts of blood elements are combined into ratios that are more robustly predicative of cancer prognosis. The neutrophil-lymphocyte ratio (NLR), the lymphocyte-monocyte ratio (LMR), and the platelet-lymphocyte ratio (PLR) have all been associated with cancer outcome (4–6). These cellular measures of inflammation can also be utilized in conjunction with protein indicators, as is the case for the prognostic index (PI) which combines CRP levels and WBC counts (7, 8).

Blood-based inflammatory biomarkers are some of the most widely used tools in assessing cancer prognosis. In addition to being easily obtainable and inexpensive to assess, they are applicable across a range of different cancer types. This chapter aims to summarize studies which assessed the ability of these inflammatory blood biomarkers to predict cancer prognosis.

COMPREHENSIVE LITERATURE SEARCH

A comprehensive literature search was conducted in Ovid MEDLINE (R) spanning 1946 to June Week 1 2016 and Embase Classic + Embase spanning 1947 to 2016 Week 24. Search terms included ‘neoplasm,’ ‘inflammation,’ ‘biomarker,’ ‘allele,’ and ‘genomics.’ The search was limited to English and human subjects.

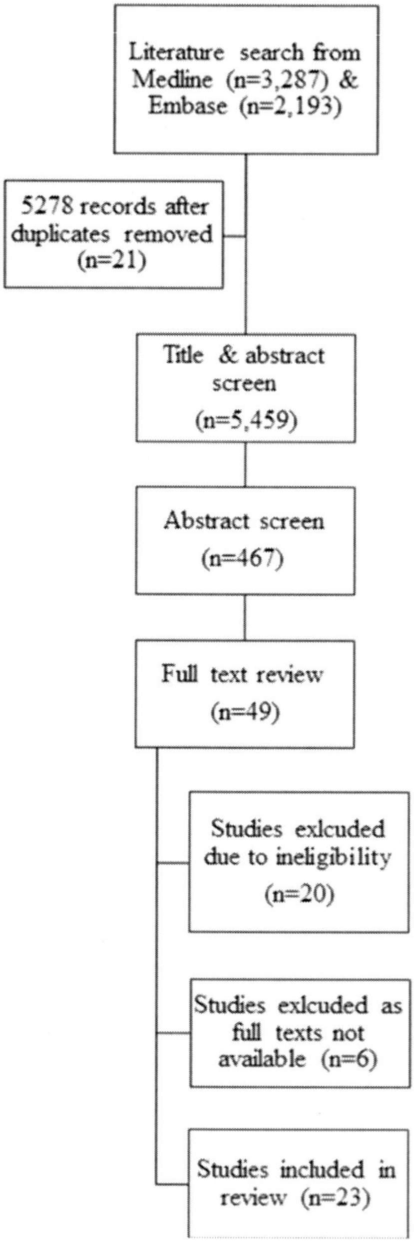

Figure 1. PRISMA flow diagram.

Study selection

References obtained from the literature search were subjected to title and abstract screening by three authors (AW, ML, AF). References were included if they discussed 1) cancer patients, 2) prognosis, and 3) common blood-based inflammation biomarkers as defined by CRP, albumin, total and specific white blood cell counts, and scores or ratios derived from them that contributed to prognosis. References that 1) lacked full text, 2) were review/meta-analysis/opinion/case studies 3) were animal or cell studies, or 4) assessed effects of clinical trials were excluded. For studies that met the inclusion criteria, full texts were obtained and reviewed by the 2 authors (AW, ML). If a discrepancy arose, discussion would ensure until a consensus could be reached. If no consensus could be reached input from another author (AF) was consulted and a decision on inclusion was made.

Data extraction

Information from each selected study was extracted including first author, publication year, patient sample size, cancer type, intervention, inflammatory blood biomarker, significance value in univariate and multivariate analysis, hazard ratio or odds ratio in multivariate analysis, and associated impact on prognosis. To maintain consistency, hazard ratios (HR) and odds ratios (OR) were represented comparing elevated biomarkers levels or biomarker scores/ratios with respect to non-elevated levels; therefore, the inverse of HR and OR was taken for studies that presented risk comparing non-elevated biomarker levels with respect to elevated levels. The data extracted from each article was reviewed by 2 authors (AW, ML) for accuracy and quality assurance.

OUR FINDINGS

Using our search strategy, a total of 3,287 original articles were identified using Medline, and 2,193 using Embase. A summary of the screening process can be found in Figure 1. Duplicates, reviews, case studies, conference abstracts, and articles that did not fulfill the inclusion criteria were eliminated, resulting in 23 studies in the analysis (see Table 1).

Table 1. Summary of study characteristics

Study characteristics	Number of studies	
	n	%
Cancer type		
Gastric cancer	5	22%
Colorectal cancer	3	13%
Hepatocellular carcinoma	3	13%
Pancreatic cancer	3	13%
Esophageal cancer	2	9%
Other cancer	4	17%
Multiple	3	13%
Treatment		
Surgery	18	78%
Chemotherapy	13	57%
Radiotherapy	5	22%
Other	2	9%
Not specified	2	9%
Multiple	11	48%
Biomarker type		
Protein-based	21	91%
Protein level	11	48%
Protein level score/ratio	15	65%
Cell based	13	57%
Cell count	3	13%
Cell count ratio	12	52%
Protein and cell based	1	4%
Inflammatory score (IS)	1	4%
Survival		
Overall survival (OS)	18	78%
Cancer specific survival (CSS)	6	26%
Disease-free survival (DFS)	5	22%
Progression-free survival (PFS)	1	4%
Distant metastasis-free survival (DMFS)	1	4%
Multiple	7	30%
Total studies	**23**	

Of the 23 articles, study sample sizes ranged from 62 to 1,825 patients, with a median sample size of 254 patients (5, 8, 10–30). The most common treatment was surgery (n=18), followed by chemotherapy (n=13) (5, 8, 10–26, 28, 30). Eleven distinct cancer types were studied, with studies on gastric cancer the most numerous (n=5), followed by colorectal cancer, hepatocellular carcinoma, and pancreatic cancer with 3 studies each (5, 8, 10–30). Prognosis was classified into five different types with overall survival (OS) being the most prevalent (n=18) (5, 8, 10–30). Other outcomes included cancer specific survival (CSS, n=6), disease free survival (DFS, n=6), progression free survival (PFS, n=1), and distant metastasis free survival (DMFS, n=1) (5, 13, 15–19, 23, 25, 28, 30). Seven studies assessed more than one type of outcome (5, 13, 16–18, 28). Prognostic inflammatory blood biomarkers were classified as protein-based (n=21), cell-based (n=13), or both (n=1), and are independently summarized in Tables 2 and 3 (5, 8, 10–30).

Studies which found significant associations between survival and protein-based inflammatory blood biomarkers derived from CRP and albumin were described in Table 2 (5, 8, 10–28). Six studies showed that high levels of albumin were associated with poor OS (n=6), DFS (n=1), or PFS (n=1) in ovarian cancer (n=1), hepatocellular carcinoma (HCC, n=2), pancreatic cancer (n=1), and gastric cancer (n=2) (5, 10–14). Eleven studies showed that high levels of CRP were associated with poor OS (n=8), CSS (n=3), DFS (n=3), PFS (n=1), or DMFS (n=1) (5, 10–13, 15–18). Four out of six associations retained significance in multivariate analysis, with respective hazard ratios (HR) ranging from 1.54 to 4.27 (11, 13, 17–19).

Table 2 also describes combined measurements of CRP and albumin through the GPS, mGPS scores, or CRP/Albumin ratio. The most common combined measure was the mGPS (n=10), followed by the GPS (n=7), and finally the CRP/Albumin ratio (n=2) (8, 10–14, 20–28). Multivariate analysis of combined measures was conducted in 17 associations, all of which returned significance, with hazard ratios (HR) that ranged from 1.41 to 7.24 (8, 10–14, 20–28). Notably, the ‘new mGPS’ introduced by Nakamura et al. had the second highest HR at 4.47 (25). Nakamura et al. observed that the existing GPS/mGPS failed to stratify patients into different prognostic groups, as the vast majority of patients had a score of 1-2, and very rarely score of ‘0’ (25). Therefore, they investigated how a change in threshold values in mGPS can result in a more sensitive system, and determined that a CRP cutoff value from 1mg/L to 0.75mg/L can more effectively stratify esophageal cancer patients (25).

Table 2. Studies that assessed CRP, albumin, and their derivative evaluation methods with outcome in cancer

Study	n	Disease	Intervention	Outcome	Significance in univariate analysis	Significant in multivariate analysis
Albumin: low levels associated with poor outcome						
Zhang et al. 2015 (9)	190	ovarian	surgery, chemotherapy	PFS OS	p<0.001 p<0.001	Not assessed Not assessed
Kinoshita et al. 2014 (10)	186	HCC	surgery, chemotherapy, other	OS	p<0.0001	Not assessed
Martin et al. 2014 (11)	124	pancreatic	chemotherapy, radiotherapy	OS	p<0.001	HR: 0.47, p<0.001
Kinoshita et al. 2013 (12)	150	HCC	surgery, chemotherapy	OS	p<0.0001	Not assessed
Wang et al. 2012 (13)	324	gastric	surgery	DFS OS	p<0.010 p<0.022	Not assessed Not assessed
Jiang et al. 2012 (14)	1710	gastric	surgery	OS	p<0.001	Not assessed
C-reactive protein (CRP): high levels associated with poor outcome						
Zhang et al. 2015 (9)	190	ovarian	surgery, chemotherapy	PFS OS	p=0.011 p=0.036	Not assessed Not assessed
Deng et al. 2015 (15)	389	gastric	surgery	CSS	p<0.013	Not assessed
Kinoshita et al. 2014 (10)	186	HCC	surgery, chemotherapy, other	OS	p<0.0001	Not assessed
Martin et al. 2014 (11)	124	pancreatic	chemotherapy, radiotherapy	OS	p=0.02	Not significant
Na et al. 2014 (16)	224	HCC	surgery, other	DFS	p=0.010	Not assessed
Szkandera et al. 2013 (17)	304	STS	surgery, radiotherapy, chemotherapy	CSS	p<0.001	HR: 2.25, p=0.010

Study	n	Disease	Intervention	Outcome	Significance in univariate analysis	Significant in multivariate analysis
				DFS	p<0.001	HR: 1.97, p=0.017
				OS	p<0.001	HR: 1.54, p=0.047
Kinoshita et al. 2013 (12)	150	HCC	surgery, chemotherapy	OS	p<0.0001	Not assessed
Xia et al. 2013 (18)	355	NPC	chemotherapy, radiotherapy	DMFS OS	p=0.001 Not assessed	p=0.007 p=0.026
Wang et al. 2012 (13)	324	Gastric	surgery	DFS	p<0.001	Not significant
				OS	p<0.001	Not significant
Jiang et al. 2012 (14)	1710	gastric	surgery	OS	p<0.001	Not assessed
Canna et al. 2004 (19)	147	CRC	surgery	CSS	p<0.001	HR: 4.27, p<0.001
Glasgow Prognostic Score (GPS): high score (high CRP, low albumin) associated with poor outcome						
Shibutani et al. 2015 (20)	254	CRC	surgery, chemotherapy	OS	p=0.004	HR: 7.238, p=0.032
Xu et al. 2015 (21)	468	esophageal	surgery, chemotherapy, radiotherapy	OS	p=0.008	Not assessed
Kinoshita et al. 2014 (10)	186	HCC	surgery, chemotherapy, other	OS	p<0.0001	Not assessed
Kinoshita et al. 2013 (12)	150	HCC	surgery, chemotherapy	OS	p<0.0001	HR: 1.773, p=0.002
Oshiro et al. 2013 (22)	62	CCC	surgery	OS	p=0.008	HR: 2.787, p=0.022
Mohri et al. 2012 (23)	197	Gastric	surgery	DFS	p<0.0001	HR: 1.81, p=0.0158

Table 2. (Continued)

Study	n	Disease	Intervention	Outcome	Significance in univariate analysis	Significant in multivariate analysis
Wang et al. 2012 (13)	324	Gastric	surgery	DFS OS	p<0.001 p<0.001	HR: 1.429, p=0.003 HR: 1.397 p=0.014
Modified Glasgow Prognostic Score (mGPS): high score (high CRP, low albumin) associated with poor outcome						
Tan et al. 2015 (24)	114	multiple	chemotherapy	OS	p<0.014	HR: 1.68, p=0.039
Xu et al. 2015 (21)	468	esophageal	surgery, radiotherapy, chemotherapy	OS	p=0.008	Not assessed
Kinoshita et al. 2014 (10)	186	HCC	surgery, chemotherapy, other	OS	p<0.0001	Not assessed
Martin et al. 2014 (11)	124	pancreatic	chemotherapy, radiotherapy	OS	p<0.001	HR: 1.41, p=0.01
Aurello et al. 2014 (8)	102	gastric	surgery, chemotherapy	DFS	p<0.001	HR: 1.78, p<0.002
				OS	p<0.001	HR: 1.70, p<0.005
Nakamura et al. 2014 (25)	168	esophageal	surgery, chemotherapy, radiotherapy	CSS	new mGPS p<0.0001 mGPS p<0.0001	HR: 4.437, p= 0.0002 HR:2.726, p=0.0449
Laird et al. 2013 (26)	1825	multiple	chemotherapy	OS	p<0.001	HR:2.05, p<0.001
Partridge et al. 2012 (27)	296	multiple	not specified	OS	not specified	HR:2.712, p=0.014
Jiang et al. 2012 (14)	1710	gastric	surgery	OS	p<0.001	OR:1.84, p<0.007

Study	n	Disease	Intervention	Outcome	Significance in univariate analysis	Significant in multivariate analysis
Mcmillan et al. 2007 (28)	316	CRC	surgery	CSS	p<0.0001	HR:1.74, p=0.0032
				OS	p<0.0001	Not assessed
CRP/Albumin Ratio: high score associated with poor outcome						
Kinoshita et al. 2014 (10)	186	HCC	surgery, chemotherapy, other	OS	p<0.0001	HR: 3.394, p=0.0001
Xu et al. 2015 (21)	468	esophageal	surgery, radiotherapy, chemotherapy	OS	p<0.001	HR: 2.44, p<0.0001

Table 3. Studies that assessed counts and ratios of platelets and leukocytes with outcome in cancer

Study	n	Disease	Intervention	Outcome	Significance in univariate analysis	Significant in multivariate analysis
Absolute neutrophil count (ANC): High cell count associated with poor outcome						
Qi et al. 2015 (29)	321	pancreatic	not specified	OS	p<0.001	HR: 2.374, p<0.001
Martin et al. 2014 (11)	124	pancreatic	chemotherapy, radiotherapy	OS	p<0.001	HR: 1.14, p<0.01
Xia et al. 2013 (18)	355	NPC	chemotherapy, radiotherapy	DMFS	p=0.042	Not significant
Absolute monocyte count (AMC): High cell count associated with poor outcome						
Qi et al. 2015 (29)	321	pancreatic	not specified	OS	p<0.001	HR: 1.725, p< 0.002
White blood cell count (WBC): High cell count generally associated with poor outcome						
Qi et al. 2015 (29)	321	pancreatic	not specified	OS	p<0.001	HR: 1.766, p< 0.001

Table 3. (Continued)

Study	n	Disease	Intervention	Outcome	Significance in univariate analysis	Significant in multivariate analysis
Xia et al. 2013 (18)	355	NPC	chemotherapy, radiotherapy	DMFS	p=0.032	Not significant
Neutrophil-Lymphocyte ratio (NLR): High ratio (high neutrophils, low lymphocytes) associated with poor outcome						
Xu et al. 2015 (21)	468	esophageal	surgery, radiotherapy, chemotherapy	OS	p=0.001	Not assessed
Zhang et al. 2015 (9)	190	ovarian	surgery, chemotherapy	PFS OS	p<0.001 p<0.001	Not assessed Not assessed
Tan et al. 2015 (24)	114	multiple	chemotherapy	OS	p<0.001	Not assessed
Deng et al. 2015 (15)	389	gastric	surgery	CSS	p=0.000	HR: 1.53, p=0.010
				DFS	p=0.000	HR: 1.38, p=0.025
				OS	p=0.000	Not assessed
Shibutani et al. 2015 (20)	254	CRC	surgery, chemotherapy	OS	Post-op NLR p=0.001 Pre-op NLR p=0.044	HR: 15.713, p=0.018 HR: 6.599, p=0.059
Qi et al. 2015 (29)	321	pancreatic	not specified	OS	p<0.001	HR: 2.033, p<0.001
Kinoshita et al. 2014 (10)	186	HCC	surgery, chemotherapy, other	OS	p=0.018	Not assessed
Martin et al. 2014 (11)	124	pancreatic	chemotherapy, radiotherapy	OS	p<0.001	HR: 1.60 p=0.02
Aurello et al. 2004 (8)	102	gastric	surgery, chemotherapy	DFS	p<0.001	Not significant
				OS	p=0.003	Not significant
Na et al. 2014 (16)	224	HCC	surgery, other	DFS	p=0.003	Not significant
				OS	p=0.003	HR: 2.896, p=0.004

Study	n	Disease	Intervention	Outcome	Significance in univariate analysis	Significant in multivariate analysis
Szkandera et al. 2013 (30)	474	pancreatic	surgery, chemotherapy	CSS	p<0.001	HR: 1.24, p=0.041
Wang et al. 2012 (13)	324	gastric	surgery	DFS	p=0.014	Not significant
				OS	p=0.013	Not significant
Platelet-Lymphocyte ratio (PLR): high ratio (high platelets, low lymphocytes) generally associated with poor outcome						
Zhang et al. 2015 (9)	190	ovarian	surgery, chemotherapy	PFS OS	p<0.001 p<0.001	HR: 1.852, p=0.001 HR: 2.158, p<0.001
Deng et al. 2015 (15)	389	gastric	surgery	CSS DFS OS	p<0.001 p<0.003 p<0.001	Not significant Not significant Not significant
Qi et al. 2015 (29)	321	pancreatic	not specified	OS	p=0.008	HR: 1.572, p<0.006
Martin et al. 2014 (11)	124	pancreatic	chemotherapy, radiotherapy	OS	p=0.01	HR: 1.58, p=0.02
Lymphocyte-Monocyte ratio (LMR): Low ratio (high lymphocytes, low monocytes) generally associated with poor outcome						
Deng et al. 2015 (15)	389	gastric	surgery	CSS DFS OS	p=0.000 p=0.000 p=0.000	Not significant Not significant Not significant
Qi et al. 2015 (29)	321	pancreatic	not specified	OS	p<0.001	HR: 0.576, p<0.001

Table 3 summarizes studies that found significant associations between cellular blood markers of inflammation. Absolute counts of white blood cell types were uncommon; only three studies assessed absolute neutrophil count (ANC), absolute monocyte count (AMC) or white blood cell count (WBC) (18, 29, 31). Multivariate analysis was performed in all three studies. Improved pancreatic cancer OS was significantly associated with high ANC (HR 1.14-2.374), AMC (HR 1.725), and WBC (HR 1.766) (18, 29, 31). However, multivariate analysis by Xia et al. on DFMS of nasopharyngeal cancer (NPC) did not find a significant association between DMFS with ANC or WBC (18).

The neutrophil-lymphocyte ratio (NLR) was the most commonly assessed out of the cellular-based biomarkers, with 12 studies that associated a high ratio with poor OS (n=11), CSS (n=1), DFS (n=3), or PFS (n=1) (5, 8, 10, 11, 13, 15, 16, 21, 24, 29, 30). NLR was studied across at least six different cancers, ranging from gastric (n=3), pancreatic (n=3), HCC (n=2), esophageal (n=1), ovarian (n=1), CRC (n=1), as well as one study by Tan et al. that assessed patients with advanced disease of different cancer types (5, 8, 10, 11, 13, 15, 16, 21, 24, 29, 30).

Multivariate analysis in eight studies on the prognostic value of NLR found that eight out of 13 associations showed that NLR was independently associated with OS, DFS, PFS or DFS (HR 1.24-15.713) (8, 11, 13, 15, 16, 20, 29, 30). Three of these were studies on gastric cancer patients who underwent surgical resection, but only one study by Deng et al. found significance in multivariate analysis (8, 13, 15). Both Aurello et al. and Wang et al. found no significance between NLR with DFS and OS at the multivariate level (8, 13). On the other hand, all three studies on pancreatic cancer found that a significant association between high NLR and poor OS or CSS was retained in multivariate analysis (HR 1.24-2.033) (11, 29, 17). Of the two studies associating NLR with prognosis of HCC, only the study by Na et al. performed multivariate analysis (16, 10). They found a significant association between NLR and OS but not with DFS in HCC patients who underwent liver transplantation (16).

The four studies that evaluated the association between high platelet-lymphocyte ratio (PLR) and poor OS (n=4), DFS (n=1), PFS (n=1), or CSS (n=1) in patients with pancreatic, ovarian, or gastric cancer. All studies performed multivariate analysis, but three out of four studies reported significant associations between high PLR and poor OS (HR 1.572-1.58) (5, 11, 29, 15). On the other hand, Deng et al. analyzed CSS, DFS, and OS of gastric cancer patients and found no significance in multivariate analysis (15).

Two studies assessed for lymphocyte-monocyte ratio (LMR), showing that low LMR is associated with poor OS (n=2), CFS (n=1), and DFS (n=1) (5, 11,

15, 29). Results from multivariate analysis was only significant in the study by Qi et al. on OS for patients with pancreatic cancer (HR 0.576), while Deng et al. found no significance for gastric cancer (29, 15).

One study by Aurello et al. evaluated the prognostic index (PI), a combined protein/cell inflammatory system that gives high scores to high CRP concentrations and high WBC count (8). Analysis of 102 patients with gastric cancer who underwent surgery found that high PI scores were significantly associated with both worse OS (HR 2.702, $p<0.001$) and DFS (HR 4.76, $p<0.001$) (8).

DISCUSSION

The abundance of literature that found significant associations between inflammatory blood biomarkers and prognosis in cancer patients supported the role of inflammation as an important factor in cancer outcomes. Of the leukocyte and platelet measures, the NLR ratio was the most often used prognostic tool and was found to be associated with outcome of at least six cancer types. Measurements of just two blood proteins, CRP and Albumin, could be used to assess prognosis in at least eight cancer types using four different methods of evaluation: protein concentrations taken at face value, and three versions of GPS scoring (GPS, mGPS, new-mGPS). The different versions of GPS were developed in an effort to improve the ability to stratify patients into different prognostic groups. However, it should be noted that a consequence of multiple evaluation methods is a possible lack of standardized reporting practice among different healthcare practitioners. Therefore, more studies comparing the methods of using CRP and albumin as a prognostic marker should be performed across a variety of cancers in order to determine an optimal, standardized scoring system.

Multivariate analysis showed variability on whether individual biomarkers or evaluation scores were independently associated with outcome. Insignificant findings for multivariate analysis were the result of six studies, of which three were performed for gastric cancer patients (8, 11, 18, 13, 15, 16). In the studies by Wang et al. and Martin et al., either CRP or albumin alone were not independent predictors of DFS or OS, but once combined together and evaluated as a GPS/mGPS, significance was retained (11, 13). This highlights the usefulness of a combined score compared to assessing the absolute levels of individual biomarkers.

Three studies found that NLR was insignificant in multivariate analysis, two of which were for gastric cancer, and one for HCC (8, 13, 16). Two out of three studies on NLR and gastric cancer found no significance at the multivariate level (8, 13, 15). The high representation of insignificance of this cancer type suggests that further study is warranted to determine the usefulness of NLR in predicting survival or recurrence in gastric cancer, and whether there are differences in NLR of gastric cancer compared to other cancer types.

The widespread correlation of high levels of inflammation-based biomarkers with cancer prognosis suggests investigation into whether inflammation itself may be a target of therapy, with anti-inflammatories given in conjunction with traditional anticancer therapy. Trials of the non-steroidal anti-inflammatory drug aspirin (NSAID) reported both significant and insignificant associations with reduced risk of colon cancer and breast cancer (33). Furthermore, a study on familial adenomatous polyposis patients found that taking the NSAID sulindac may decrease polyp number and cancer recurrence (34). As increased inflammation is associated with many cancer types, the combined targeting of inflammation and cancer may lead to increased treatment efficacy, potentially across varied patient populations.

The inflammation-based prognostic scores and markers discussed have proven suggestive of the likely course of disease. These prognostic blood biomarkers are not only simple and economical to test for, but also provide prognostic information on duration of expected survival. This information is invaluable for patients, family members, and healthcare practitioners in order to help them prepare and plan for the future, such as switching from treatment to supportive or palliative care. Therefore, the ability of the blood-based inflammatory compounds to predict the outcome of many cancers can become a guideline for deciding the most appropriate form of care.

ACKNOWLEDGMENT

We thank the generous support of Joey and Mary Furfari Cancer Research Fund.

REFERENCES

[1] Hanahan D, Weinberg RA. Hallmarks of cancer: The next generation. Cell 2011;144(5):646–74.

[2] Rakoff-Nahoum S. Why cancer and inflammation? Yale J Biol Med 2006;79(3-4):123–30.

[3] McMillan DC. The systemic inflammation-based Glasgow Prognostic Score: A decade of experience in patients with cancer. Cancer Treat Rev 2013;39(5):534–40.

[4] Templeton AJ, Ace O, McNamara MG, Al-Mubarak M, Vera-Badillo FE, Hermanns T, et al. Prognostic Role of platelet to lymphocyte ratio in solid tumors: a systematic review and meta-analysis. Cancer Epidemiol Biomarkers Prev 2014;23(7):1204–12.

[5] Zhang W, Liu K, Hu G, Liang W. Preoperative platelet/lymphocyte ratio is a superior prognostic factor compared to other systemic inflammatory response markers in ovarian cancer patients. Tumor Biol 2015;36(11):8831–7.

[6] Hu P, Shen H, Wang G, Zhang P, Liu Q, Du J. Prognostic significance of systemic inflammation-based lymphocyte- monocyte ratio in patients with lung cancer: based on a large cohort study. PLoS One 2014;9(9):e108062.

[7] Pinato DJ, North B V, Sharma R. A novel, externally validated inflammation-based prognostic algorithm in hepatocellular carcinoma: the prognostic nutritional index (PNI). Br J Cancer 2012;106(8):1439–45.

[8] Aurello P, Tierno SM, Berardi G, Tomassini F, Magistri P, D'Angelo F, et al. Value of preoperative inflammation-based prognostic scores in predicting overall survival and disease-free survival in patients with gastric cancer. Ann Surg Oncol 2014;21(6):1998–2004.

[9] Zhang WW, Liu KJ, Hu GL, Liang WJ. Preoperative platelet/lymphocyte ratio is a superior prognostic factor compared to other systemic inflammatory response markers in ovarian cancer patients. Tumour Biol 2015;36(11):8831–7.

[10] Kinoshita A, Onoda H, Imai N, Iwaku A, Oishi M, Tanaka K, et al. The C-reactive protein/albumin ratio, a novel inflammation-based prognostic score, predicts outcomes in patients with hepatocellular carcinoma. Ann Surg Oncol 2014;22(3):803–10.

[11] Martin HL, Ohara K, Kiberu A, Van Hagen T, Davidson A, Khattak MA. Prognostic value of systemic inflammation-based markers in advanced pancreatic cancer. Intern Med J 2014;44(7):676–82.

[12] Kinoshita A, Onoda H, Imai N, Iwaku A, Oishi M, Tanaka K, et al. The Glasgow Prognostic Score, an inflammation based prognostic score, predicts survival in patients with hepatocellular carcinoma. BMC Cancer 2013;13:52.

[13] Wang DS, Ren C, Qiu MZ, Luo HY, Wang ZQ, Zhang DS, et al. Comparison of the prognostic value of various preoperative inflammation-based factors in patients with stage III gastric cancer. Tumor Biol 2012;33(3):749–56.

[14] Jiang X, Hiki N, Nunobe S, Kumagai K, Kubota T, Aikou S, et al. Prognostic importance of the inflammation-based Glasgow prognostic score in patients with gastric cancer. Br J Cancer 2012;107(2):275–9.

[15] Deng Q, He B, Liu X, Yue J, Ying H, Pan Y, et al. Prognostic value of pre-operative inflammatory response biomarkers in gastric cancer patients and the construction of a predictive model. J Transl Med 2015;13:66.

[16] Na GH, Kim DG, Han JH, Kim EY, Lee SH, Hong TH, et al. Inflammatory markers as selection criteria of hepatocellular carcinoma in living-donor liver transplantation. World J Gastroenterol 2014;20(21):6594–601.

[17] Szkandera J, Gerger A, Liegl-Atzwanger B, Absenger G, Stotz M, Samonigg H, et al. Validation of the prognostic relevance of plasma C-reactive protein levels in soft-tissue sarcoma patients. Br J Cancer 2013;109(9):2316–22.

[18] Xia WX, Zhang HB, Shi JL, Lu X, Wang L, Ye YF, et al. A prognostic model predicts the risk of distant metastasis and death for patients with nasopharyngeal carcinoma based on pre-treatment serum C-reactive protein and N-classification. Eur J Cancer 2013;49(9):2152–60.

[19] Canna K, McMillan DC, McKee RF, McNicol A-M, Horgan PG, McArdle CS. Evaluation of a cumulative prognostic score based on the systemic inflammatory response in patients undergoing potentially curative surgery for colorectal cancer. Br J Cancer 2004;90(9):1707–9.

[20] Shibutani M, Maeda K, Nagahara H, Ohtani H, Iseki Y, Ikeya T, et al. The prognostic significance of a postoperative systemic inflammatory response in patients with colorectal cancer. World J Surg Oncol 2015;13:194.

[21] Xu XL, Yu HQ, Hu W, Song Q, Mao WM. A novel inflammation-based prognostic score, the c-reactive protein/albumin ratio predicts the prognosis of patients with operable esophageal squamous cell carcinoma. PLoS One 2015;10(9):e0138657.

[22] Oshiro Y, Sasaki R, Fukunaga K, Kondo T, Oda T, Takahashi H, et al. Inflammation-based prognostic score is a useful predictor of postoperative outcome in patients with extrahepatic cholangiocarcinoma. J Hepatobiliary Pancreat Sci 2013;20(3):389–95.

[23] Mohri Y, Tanaka K, Ohi M, Toiyama Y, Yasuda H, Inoue Y, et al. Inflammation-based prognostic score as a predictor of postoperative gastric cancer recurrence. Anticancer Res 2012;32(10):4581–4.

[24] Tan CSY, Read JA, Phan VH, Beale PJ, Peat JK, Clarke SJ. The relationship between nutritional status, inflammatory markers and survival in patients with advanced cancer: a prospective cohort study. Support Care Cancer 2015;23(2):385–91.

[25] Nakamura M, Iwahashi M, Nakamori M, Ojima T, Katsuda M, Iida T, et al. A new prognostic score for the survival of patients with esophageal squamous cell carcinoma. Surg Today 2014;44(5):875–83.

[26] Laird BJ, Kaasa S, McMillan DC, Fallon MT, Hjermstad MJ, Fayers P, et al. Prognostic factors in patients with advanced cancer: a comparison of clinicopathological factors and the development of an inflammation-based prognostic system. Clin Cancer Res 2013;19(19):5456–64.

[27] Partridge M, Fallon M, Bray C, McMillan D, Brown D, Laird B. Prognostication in advanced cancer: A study examining an inflammation-based score. J Pain Symptom Manage 2012;44(2):161–7.

[28] McMillan DC, Crozier JEM, Canna K, Angerson WJ, McArdle CS. Evaluation of an inflammation-based prognostic score (GPS) in patients undergoing resection for colon and rectal cancer. Int J Colorectal Dis 2007;22(8):881–6.

[29] Qi Q, Lee IK, Lee YS, Sung NY, Kwon TS. Clinical implications of systemic inflammatory response markers as independent prognostic factors in colorectal cancer patients. Pancreatology 2015;15(2):145–50.

[30] Szkandera J, Stotz M, Eisner F, Absenger G, Stojakovic T, Samonigg H, et al. External validation of the derived neutrophil to lymphocyte ratio as a prognostic marker on a large cohort of pancreatic cancer patients. PLoS One 2013;8(11):e78225.

[31] Ross RW, Galsky MD, Scher HI, Magidson J, Magidson K, Lee GSM, et al. A whole-blood RNA transcript-based prognostic model in men with castration-resistant prostate cancer: A prospective study. Lancet Oncol 2012;13(11):1105–13.

[32] Pu X, Hildebrandt M, Charles L, Roth J, Stewart D, Yang Z, et al. Inflammation-related genetic variations and survival for advanced non-small cell lung cancer receiving first-line chemotherapy. Clin Pharmacol Ther 2014;96(3):360–9.

[33] Rayburn ER, Ezell SJ, Zhang R. Anti-inflammatory agents for cancer therapy. Mol Cell 2009;1(1):29–43.

[34] Kim KY, Jeon SW, Park JG, Yu CH, Jang SY, Lee JK, et al. Regression of colonic adenomas after treatment with sulindac in familial adenomatous polyposis: A case with a 2-year follow-up without a prophylactic colectomy. Ann Coloproctol 2014;30(4):201–4.

In: Oncology
Editors: A. Furfari et al.
ISBN: 978-1-53610-608-4

Chapter 9

BIOLOGICAL BIOMARKERS OF CANCER PROGNOSIS

Bo Angela Wan[1], MPhil, Michael Lam[1], BMSc(C), Anthony Furfari[1], MD(C), Rachel McDonald[1], MD(C), Leigha Rowbottom[1], MD(C), Stephanie Chan[1], BSc(C), Pearl Zaki[1], BSc(C), Carlo DeAngelis[1], PharmD, Ronald Chow[1], BMSc(C), George S Charames[2,3,4], PhD, Henry Lam[1], MLS, Edward Chow[1], MBBS and Azar Azad[2,3,5,*], PhD

[1]Odette Cancer Centre, Sunnybrook Health Sciences Centre, University of Toronto, Toronto, Ontario, Canada
[2]Pathology and Laboratory Medicine, Mount Sinai Hospital, Toronto, Ontario, Canada
[3]Laboratory Medicine and Pathobiology, University of Toronto, Toronto, Ontario, Canada
[4]Lunenfeld-Tanenbaum Research Institute, Sinai Health System, Toronto, Ontario, Canada
[5]Mount Sinai Services Inc., Toronto, Ontario, Canada

* Correspondence: Azar Azad, PhD, FACMG, Pathology and Laboratory Medicine, Mount Sinai Hospital, Suite 6-500, 600 University Ave, Toronto, Ontario, M5G 1X5, Canada. E-mail: aazad@mountsinaiservices.com.

Antibody-based detection methods have allowed the evaluation of many biological compounds of clinical importance in cancer prognosis. These include proteins in bodily fluids, proteins in tumor tissue or lymph nodes, and cell types identifiable by protein expression. In this chapter we systematically summarized the literature on uncommon molecular biomarkers associated with cancer outcome. A literature search was conducted on Medline and Embase utilizing keywords such as "cancer," "inflammation," "prognosis," "inflammation" and "biomarkers." Articles that reported on uncommon biological biomarkers of prognosis were selected for inclusion. Information regarding the cancer type, interventions received, and the impact on prognosis and significance was extracted. A total of 28 studies comprising of 5,234 patients found associations between 45 uncommon molecular biomarkers and prognosis in more than 15 types of primary malignancies. Many molecular biomarkers found to be associated with prognosis are part of the overlapping pathways of inflammation and oxidative stress. Identification of prognostic biomarkers has the potential to highlight aspects of physiology that lead to poor prognosis and assist in the identification of therapeutic targets.

INTRODUCTION

Cancer prognosis is important to patients and physicians alike, as knowing the likely course of disease can allow the appropriate psychological and medical preparation. Prognosis has traditionally been based on cancer grading, whereby a semi-quantitative combination of the patients' physiology and cancer characteristics are evaluated into a classification of cancer severity, the degree of which is indicative of survival (1, 2). Prognostic tests that evaluate for certain bodily compounds have also been developed, such as the assessment of blood biomarkers albumin, CRP, and leukocyte levels that are indicators of systemic inflammation. These are among the most well-established indicators of outcome in multiple cancer types, and are separately discussed in a chapter within this publication. While blood biomarkers are widely applicable in predicting overall survival in cancer, additional biological biomarkers are needed to more accurately predict prognosis and also to define specific cancer characteristics. Therefore, there is an ongoing effort to discover additional molecular biological biomarkers for their significance in predicting survival, in addition to providing potential targets for research into cancer management and treatment.

What may begin as a prognostic marker may in fact be an indicator of a vital aspect of cancer biology, and lead to the development of a targeted therapy. The iconic example of one of the most successful results of this approach is the human epidermal growth factor receptor 2 (HER2) biomarker for breast cancer, which predicts response to HER2 therapy in addition to being a prognostic biomarker (2). Regular assessment of the HER2 biomarker status and clinical implementation of trastuzumab, the monoclonal antibody against HER2, has enabled a drastic improvement in survival of HER2-positive breast cancer patients (2). In addition to potentially introducing treatment options for stratified populations of cancer patients, prognostic biomarkers have value in enabling effective decision-making for patients and physicians. Knowing the likely course of disease can inform the goals of care, such as implementing advanced directives or switching from a curative to palliative approach. Accurate estimates of prognosis are required to ensure that patients are maximizing quality of life (QOL) while minimizing adverse events towards the end of life.

Our review focuses on summarizing the literature on cell or protein biomarkers, in particular, uncommon prognostic molecular biomarkers in cancer that are associated with outcome in cancer patients. Outcomes of interest included overall survival (OS), cancer-specific survival (CSS), disease-free survival (DFS), and recurrence-free survival (RFS).

Our literature review

A comprehensive literature search was conducted in Ovid MEDLINE (R) spanning 1946 to June Week 1 2016 and Embase Classic + Embase spanning 1947 to 2016 week 24. Search terms included 'neoplasm', 'inflammation', 'biomarker', 'allele', and 'genomics'. The search was limited to English and human subjects (see Figure 1).

Study selection

Two authors (AW, ML) screened the titles and abstracts from all references obtained from the literature search. Articles were included if they discussed all of the following topics: 1) cancer patients, 2) prognosis, and 3) uncommon biological biomarkers. References that 1) lacked full text, 2) were review/meta-analysis/opinion/case studies or 3) assessed effects of clinical trials were excluded. For studies that met the inclusion criteria, full texts were obtained and reviewed by two authors (AW, ML). If a discrepancy arose, discussion ensued until a consensus was reached. If no consensus was reached, another author (AF) was consulted to make the final decision on inclusion.

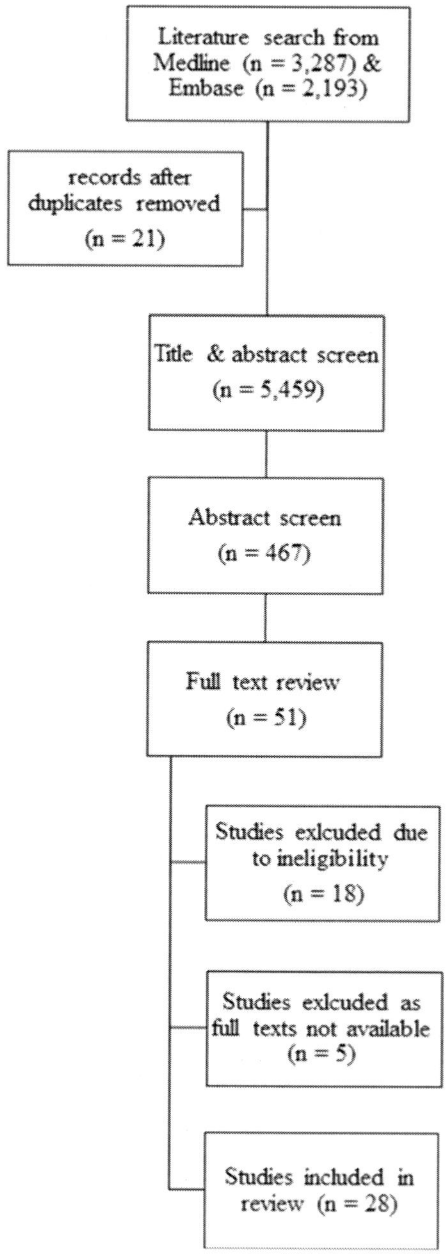

Figure 1. PRISMA flow diagram.

Table 1. Summary of study characteristics

Study characteristics	Number of studies	
	n	**%**
Cancer type		
Lung cancer	7	25%
Colorectal cancer (CRC)	4	14%
Gastric cancer	3	11%
Malignant fibrous hitsocytoma (MFH)	2	7%
Bladder cancer	2	7%
Prostate	2	7%
Other cancer	7	25%
Multiple	1	4%
Treatment		
Surgery	23	82%
Chemotherapy	7	25%
Radiotherapy	3	11%
Not specified	1	4%
Multiple	7	25%
Biomarker type		
Blood protein	12	43%
Pleural effusion protein	1	4%
Tissue molecule	12	43%
Tissue cell type	7	25%
Survival		
Overall survival (OS)	22	79%
Cancer specific survival (CSS)	5	18%
Disease-free survival (DFS)	7	25%
Recurrence-free survival (RFS)	1	4%
Multiple	4	14%
Total studies	**28**	

Data extraction

The following information from each selected study was extracted: first author name, publication year, patient sample size, cancer type, intervention, significant biomarker, biomarker source, outcome measured, and univariate p-value. If multivariate analysis was performed, hazard ratios (HR), odds ratios (OR), relative risk (RR) and associated p-values were also extracted. If necessary, the inverse of the reported HR, OR, and RR was taken in order to consistently represent the risk associated with higher levels of biomarkers with respect to low/baseline levels as a reference, rather than the risk associated with low levels of biomarkers using high levels as a reference. The data extracted from each article was reviewed by two authors (AW, ML) for accuracy and quality assurance.

What we found in our search

The search returned 3,287 records from Medline and 2,193 from Embase. After elimination of duplicates and ineligible articles, 28 studies remained for analysis. The general characteristics of the included studies are summarized in Table 1. Study cohort sizes ranged from 25 to 724 patients, with a median study size of 114. The two most common interventions were surgery (n=23) and chemotherapy (n=7), with seven studies reporting multiple intervention types. Overall survival (OS) was the most prominent outcome evaluated (n=22), followed by disease-free survival (DFS, n=7), cancer-specific survival (CSS, n=5), and recurrence-free survival (RFS, n=1). Four studies assessed multiple types of outcomes.

Table 2 summarizes the 13 studies which found protein-based markers in bodily fluids that were significantly associated with prognosis in univariate analysis (3–15). Twelve studies were conducted on individual cancer types, and one study pooled multiple cancer types (3–15). Twelve studies analyzed blood samples, while one study, by Jiang et al., analyzed pleural effusion (3–15). Multivariate analyses were performed in 8 studies, together finding 12 proteins that retained significance suggesting that they are independently correlated with outcome (3–5, 7, 10, 13–15).

Four biomarkers, hemoglobin (Hb), carcinoembryonic antigen (CEA), pentraxin 3 (PTX3), and interleukin 6 (IL6), were found to be significantly associated with poor CSS, DFS, or OS, in multiple univariate analyses, and were also found to be significant in multivariate analysis in at least one study (3, 4, 9, 10, 13–15). Additional biomarkers that are independent predictors of survival include the interleukins IL1, IL10 and IL8, vascular endothelial growth factor (VEGF), and prostate-specific antigen (PSA) (3, 5, 7). In addition, multivariate analysis showed that elevated blood CEA was associated with the greatest risk of death (OR 12.5, $p<0.001$), and elevated blood IL1 was associated with the greatest risk of recurrence (HR=11.84, $p=0.01$) (3, 14).

Table 2. Significant molecular biomarkers found in bodily fluids

Study	n	Cancer type	Treatment	Biomarker	Outcome associated with high biomarker level		
					Outcome	Univariate Analysis	Multivariate Analysis
Di Caro et al., 2016 (3)	69	CRC	Surgery	**IL1** **IL10** **PTX3** **IL8** **VEGF**	shorter DFS and shorter DSS for all	DFS: p = 0.01 DFS: p = 0.01 DFS: NS; CSS: p = 0.005 DFS:p < 0.001; CSS: p = 0.008 DFS:p < 0.001; CSS: p = 0.001	DFS: HR 11.84 (p = 0.01) DFS: (p = 0.01) DFS: HR 6.94 (p = 0.002) DFS: (p < 0.001) DFS: (p < 0.001)
Infante et al., 2016 (4)	106	NSCLC	Surgery	Hb PTX3	shorter DFS and shorter OS for both	DFS: p = 0.011; OS: p = 0.013 DFS: NS; OS: p < 0.001	Not assessed
Koo et al., 2015 (5)	440	Prostate	Hormone therapy	**PSA**	shorter CSS	p < 0.001	HR 1.002 (p < 0.001)
Wei et al., 2015 (6)	100	NSCLC	Surgery	HMNG1	shorter OS	OS: p = 0.019	Not assessed
Mori et al., 2015 (7)	349	HCC	Surgery	**AFP**	shorter OS	p < 0.001	HR 1.694 (p = 0.002)
Jiang et al., 2015 (8)	116	NSCLC	Chemotherapy	ERCC1 TYMS TUBB3 Myosin II Myoglobin MyoD1	shorter OS	p = 0.000 p = 0.000 p = 0.001 p = 0.008 p = 0.000 p = 0.000	Not assessed

Table 2. (Continued)

Study	n	Cancer type	Treatment	Biomarker	Outcome associated with high biomarker level		
					Outcome	**Univariate Analysis**	**Multivariate Analysis**
Wu et al., 2014 (9)	85	Gallbladder	Surgery	CA19-9 CEA CA125	Shorter OS	$p = 0.002$ $p = 0.001$ $p < 0.001$	Not significant
Suzuki et al., 2014 (10)	112	MPM	Surgery, chemotherapy, palliation	**Hb** LD	Shorter OS	$p = 0.0018$ $p = 0.009$	HR 1.94 ($p = 0.01$) Not assessed
Menschikowski et al., 2013 (12)	73	multiple	Not specified	sPLA2-IIA	shorter OS	$p < 0.01$	Not assessed
Bodelon et al., 2013 (11)	241	NSCLC	Surgery, chemotherapy, radiotherapy	CCL15	shorter OS	$p = 0.042$	Not assessed
Son et al., 2013 (13)	642	CRC	Surgery, chemotherapy	**CEA** **Fibrinogen**	shorter DFS and OS	DFS: $p = 0.001$; OS: $p = 0.005$ DFS: $p = 0.005$; OS: $p = 0.010$	HR 1.793 ($p = 0.010$) HR 1.999 ($p = 0.027$)
Dehing-Oberije et al., 2011 (14)	158	NSCLC	Chemotherapy, radiotherapy	**CEA** IL6 OPN CYFRA21-1	shorter OS	$p = 0.013$ $p = 0.003$ $p = 0.0038$ $p < 0.001$	OR 12.5 ($p < 0.001$) Not significant Not assessed Not assessed
Alcover et al., 2010 (15)	96	Prostate	Surgery	**IL6**	shorter RFS	$p = 0.031$	HR not available, ($p = 0.04$)

* Bolded biomarkers are significant in multivariate analysis.

Table 3. Significant molecular biomarkers found in tissues

Study	n	Cancer type	Treatment	Biomarker	Outcome associated with high biomarker level		
					Outcome	Univariate Analysis	Multivariate Analysis
Infante et al., 2016 (4)	106	NSCLC	Surgery	**PTX3**	shorter DFS & shorter OS	DFS: $p < 0.006$ OS: $p < 0.001$	DFS: HR 4.387 ($p = 0.002$) OS: HR 8.534 ($p < 0.001$)
Seppälä et al., 2015 (16)	150	Tongue	Surgery	Metastasis CIP2A lymph node CIP2A	shorter DFS shorter DFS	$p = 0.000$ $p = 0.042$	Not assessed
Bergenfelz et al., 2015 (17)	144	Breast	Surgery	MRP14	shorter OS	$p = 0.001$	Not assessed
Youssef et al., 2015 (18)	151	Bladder squamous cell carcinoma	Surgery	**FGF**	shorter DFS shorter OS	$p < 0.001$ $p = 0.001$	HR 2.561 ($p = 0.009$) HR 2.679 ($p = 0.033$)
Wouters et al., 2014 (20)	160	Ovarian	Surgery, chemotherapy	IL6 **IL6R**	shorter CSS longer CSS	$p = 0.034$ $p = 0.010$	Not significant HR 0.474 ($p = 0.011$)
Zong et al., 2014 (19)	300	CRC	Surgery	DcR3	shorter OS	$p<0.001$	Not assessed
Dumitru et al., 2013 (21)	83	Laryngeal carcinoma	Surgery	**AHNAK**	shorter OS	$p = 0.021$	HR 3.17 ($p = 0.004$)
Huang et al., 2013 (22)	60	Cervical	Surgery	CXCR6	shorter OS	$p = 0.049$	Not assessed

Table 3. (Continued)

Study	n	Cancer type	Treatment	Biomarker	Outcome associated with high biomarker level		
					Outcome	Univariate Analysis	Multivariate Analysis
Morikawa et al, 2011 (23)	724	CRC	Surgery	**STAT3**	shorter CSS	$p = 0.0020$	HR 1.61 ($p = 0.015$)
Hoki et al., 2007 (24)	36	Malignant fibrous histiocytoma	Surgery	HIF1α	shorter OS	$p = 0.01104$	Not assessed
Hoki et al., 2007 (25)	25	Malignant fibrous histiocytoma	Surgery	iNOS NFκB PTGS2 (COX2)	shorter OS for all	$p = 0.0278$ $p = 0.0080$ $p = 0.0007$	Not assessed
Hilmy et al., 2006 (26)	103	Bladder transitional cell carcinoma	Surgery, radiotherapy	PTGS2 (COX2)	shorter CSS	$p < 0.05$	Not significant

* Bolded biomarkers are significant in multivariate analysis.

Table 4. Significant cellular biomarkers found in tissues

Study	n	Cancer type	Treatment	Biomarker	Outcome associated with positive biomarker/high biomarker level		
					Outcome	Univariate Analysis	Multivariate Analysis
Kim et al, 2014 (27)	457	Gastric	Surgery, chemotherapy (5-fluorouracil)	**KRS(+) Tumor cells** **KRS(+) TAI cells**	shorter OS longer OS	$p = 0.003$ $p = 0.049$	KRS(+)Tumor + KRS(-)TAI associated with shorter OS (HR2.22 $p = 0.001$)
Belmont et al, 2014 (28)	61	NSCLC	Surgery	TLR9(+) mononuclear cells	shorter OS	$p = 0.03$	Not significant
Wouters et al, 2014 (20)	160	Ovarian	Surgery, chemotherapy	CD33(-) macrophages	shorter CSS	$p = 0.017$	Not significant
Dumitru et al, 2013 (21)	83	Laryngeal carcinoma	Surgery	CD66b(+) neutrophils	shorter OS	$p = 0.048$	Not significant
Fan et al, 2012 (29)	117	Gastric	Surgery	**MRP14(+) inflammatory cells**	longer OS	$p = 0.021$	RR 0.643 $p = 0.046$
Akaike et al, 2007 (30)	76	Gastric	surgery	HMGB1(+) cancer cells	longer OS	$p < 0.05$	Not significant
Hilmy et al, 2006 (26)	103	Bladder transitional cell carcinoma	Surgery, radiotherapy	Ki67(+) tumour cells	shorter CSS	$p < 0.05$	Not significant

* Bolded biomarkers are significant in multivariate analysis.

One study by Menschikowski et al., pooled data from more than 8 types of solid cancers and two types of hematological malignancies for prognostic inflammatory and coagulation markers in blood (12). The authors found that up-regulation of secreted phospholipase A2 type IIA (sPLA2-IIA) was associated with shorter OS ($p<0.01$) in univariate analysis (12). Jiang et al. found 6 proteins in pleural effusion that were predictive of poor OS in lung cancer patients in univariate analysis (8). This included increased levels of several proteins implicated in drug resistance, including excision repair cross-complementing gene 1 (ERCC1), thymidylate synthase (TYMS), ribonucleotide reductase M1 (RRM1), and βIII-tubulin (TUBB3) (8).

Twelve studies that assessed for molecular biomarkers detected through immunohistochemistry of tissues such as excised tumour, tumour biopsy, or lymph node biopsy are summarized in Table 3 (4, 16–26). These corresponded to 10 different primary cancer types that found 14 tissue biomarkers that were significantly associated with prognosis in univariate analysis. Six studies performed multivariate analysis and found that 5 proteins were independently associated with OS, DFS, or CSS (4, 18, 20, 21, 23, 26).

Infante et al., found that high levels of PTX expression in lung tumour tissue were independently associated with short DFS and OS (4). In addition, multivariate analysis found that high levels of the transcription factor signal transducer and activator of transcription3 (STAT3), the cell migration molecule desmoyokin (AHNAK), and fibroblast growth factor (FGF) were also individually and independently associated with poor OS and CSS (18, 21, 23). In contrast, high levels of the cytokine receptor IL6R were independently associated with improved CSS (20). The risk of mortality associated with elevated levels of biomolecule compared to baseline levels was highest at 8.53 times for PTX, followed by 3.17 times for ANHAK (4, 21).

Two studies by Hoki et al., had partially overlapping study populations of malignant fibrous histiocytoma patients who underwent surgical resection at the same Japanese hospital from 1989-2004 or from 1990-2004 (24, 25). The two studies found that high levels of the hypoxia inducible factor 1α (HIF1α), and pro-inflammatory mediators such as inducible nitric oxide synthase (iNOS), nuclear-factor kappa b (NFκB), and the prostaglandin-endoperoxide synthase 2 enzyme (PTGS2) were all associated with poor OS in univariate analysis (24, 25).

Seven studies that analyzed biomarkers consisting of cells expressing specific proteins in tumour tissue are described in Table 4 (20, 21, 26–30). Univariate analysis found 8 cell types that were significantly associated with CSS or OS across 5 cancer types, including gastric cancer (n=3), ovarian cancer (n=1), non-small cell lung cancer (NSCLC) (n=1), laryngeal carcinoma (n=1) and bladder transitional cell carcinoma (n=1) (20, 21, 26–30). Cell types included immune-related cells such as macrophages or inflammatory cells (n=5), as well as tumour cells (n=3). Wouters et al. analyzed ovarian cancer tissue and found that high levels of CD33(-) macrophages were associated with improved OS in univariate analysis (p=0.034), but did not retain significance in multivariate analysis (20). Similarly assessing immune cells, Belmont et al. and Dumitru et al. respectively found that mononuclear cells expressing toll-like receptor 9 (TLR9) and neutrophils expressing CD66b were both associated with shorter OS in univariate analysis, but not in multivariate analysis (21, 28).

Multivariate analysis was conducted on all cell types and found that three were significant independent predictors of outcome (27,29). Kim et al, studied the expression of Lysyl-tRNA Synthetase on tumour cells and on tumour-associated inflammatory (TAI) cells, and found that a (KRS+) tumour, KRS(-)TAI profile was independently associated with 2.2 times increased risk of death in gastric cancer (27). Also assessing gastric cancer, Fan et al., found that high levels of inflammatory cells that express the calcium-binding protein S100A9 (also known as the migration inhibitory factor-related protein MPR14) were independently associated improved OS in gastric cancer (29).

DISCUSSION

The biomarkers described in this review are often part of recurring physiological pathways that become dysregulated in cancers. For example, proteins involved in the inflammatory response have frequently been identified as biomarkers predictive of adverse outcomes. This includes pro-inflammatory signaling molecules such as IL1, IL6, IL8, and PTX3, but also the anti-inflammatory IL10 (3, 4, 14, 20). Of particular significance is PTX3, a pro-inflammatory protein involved with the innate immune response and tissue remodeling that belongs to the same molecular class as the better known C-reactive protein (CRP). Multivariate analysis performed on three associations between blood or tissue PTX3 and outcome have all found PTX3 to be a significant independent predictor of poor DFS or OS (DFS HR = 4.39-6.94, OS HR = 8.53, $p < 0.01$ for all) (3, 4). As one study was performed on colorectal cancer (CRC) and the other on NSCLC, this suggests that PTX3 may be similar to CRP in being a marker that is broadly significant in predicting poor outcomes across a range of cancer types (3, 4).

One of the ways in which inflammation facilitates cancer growth and progression is through providing an oxidative environment. The inflammation-activated transcription factor NFκB, and the hypoxia-activated transcription factor HIF1α both lead to production of inducible nitric oxide synthase (iNOS), the enzyme that produces the free radical nitric oxide (NO) (25). The production of NO is part of the immune defence mechanism to kill parasites, bacteria, and cancer cells. Since NO damages a cell by mutating its DNA and causing genetic instability, this ironically contributes to carcinogenesis through providing the potential for cancer cells to acquire new mutations that enable growth, metastasis, and evasion of immune detection (31). Hoki et al, demonstrated that upregulation of elements in this pathway—NFκB, HIF1α, and iNOS, as well as evidence of DNA damage (8-nitroguanine, 8-oxodG) are correlated with poor OS for patients with malignant fibrous histiocytoma (24, 25). Further investigation of elements in this pathway should be conducted in other cancer types to verify its broader significance.

Angiogenesis is a third pathway related to both inflammation and oxidation that is frequently implicated in cancer progression. As mentioned above, inflammation activates the transcription factor HIF1α, which leads to the transcription of genes such as vascular endothelial growth factor (VEGF) that are involved in stimulating the formation of new blood vessels. Hoki et al, found that HIFα overexpression in malignant fibrous histiocytoma tissues was significantly associated with reduced OS in univariate analysis (24). Di Caro et al, found that high levels of VEGF in the bloodstream of patients with CRC was significantly associated with reduced DFS and CSS in both univariate and multivariate analysis (3).

The fact that elevated levels of biomarkers of angiogenesis were associated with poor prognosis corroborates the theory that angiogenesis is a hallmark of cancer by nourishing developing tumours and enabling their expansion (31). This suggests that limiting angiogenesis may also limit cancer growth (32). Developments from this research led to the introduction of bevacizumab, a monoclonal antibody that binds and inhibits VEGF and reduces blood vessel growth (32). Bevacizumab has demonstrated efficacy in multiple major cancer types, including CRC and NSCLC (33). Moreover, the research and development of antibody-based targeted therapies like bevacizumab has flourished in the past decade, and is projected to be one of the most important sources of upcoming cancer therapies (34).

Detection of the biomarkers described in this review is often facilitated by antibodies. The use of antibody-associated techniques has several merits, including their flexibility and specificity. Antibodies can be designed to bind to nearly any type of target, whether they are proteins, or non-protein organic compounds, as well as inorganic compounds. Moreover, multiple antibodies can be incorporated into a single assay in order to strengthen its specificity and sensitivity (35). Therefore, antibody-based techniques have high potential in the validation of a large variety of prognostic biomarkers, as well as in the translation into a diagnostic test for clinical use. Antibodies are now directly used in cancer treatment as part of a method known as immunotherapy that controls tumours through manipulating components of the body's own immune system. The US Food and Drug Administration (FDA) have recently approved several immune-checkpoint inhibitors, which are immunotherapy methods that utilize monoclonal antibodies to modulate immune reactions against cancer cells (36). Nivolumab and pembrolizumab are two such examples shown to be effective in advanced melanoma, NSCLC, and renal cell carcinoma (36). They function by blocking the programed cell death protein 1 pathway (anti-PD1) that is often coopted by cancer cells to evade the immune response (36).

Our review has summarized 45 unique molecular biomarkers are associated with cancer outcome. This number is likely to grow as technological equipment advances and as insight into human physiology continues to improve. Increased discoveries of significant prognostic biomarkers may allow the formation of prognostic formulas that include multiple biomarker types that are able to more accurately and more broadly predict outcome. The generation of an appropriate prognostic formula relies on in-depth and comprehensive analysis of many new biomarkers, which highlights the importance of continued research of biomarker prognosis in multiple cancer types. Ultimately, understanding the nature of biological biomarkers not only improves prognosis, but also facilitates the identification of potential interventional targets for personalized therapy.

ACKNOWLEDGMENT

We thank the generous support of Joey and Mary Furfari Cancer Research Fund.

References

[1] Chen M, Huang J, Zhu Z, Zhang J, Li K. Systematic review and meta-analysis of tumor biomarkers in predicting prognosis in esophageal cancer. BMC Cancer 2013;13:539.

[2] Weigel MT, Dowsett M. Current and emerging biomarkers in breast cancer: Prognosis and prediction. Endocr Relat Cancer 2010;17(4):R245-62.

[3] Di Caro G, Carvello M, Pesce S, Erreni M, Marchesi F, Todoric J, et al., Circulating inflammatory mediators as potential prognostic markers of human colorectal cancer. PLoS One 2016;11(2): e0148186.

[4] Infante M, Allavena P, Garlanda C, Nebuloni M, Morenghi E, Rahal D, et al., Prognostic and diagnostic potential of local and circulating levels of pentraxin 3 in lung cancer patients. Int J Cancer 2016;138(4):983–91.

[5] Koo KC, Park SU, Kim KH, Rha KH, Hong SJ, Yang SC, et al., Prognostic impacts of metastatic site and pain on progression to castrate resistance and mortality in patients with metastatic prostate cancer. Yonsei Med J 2015;56(5):1206–12.

[6] Wei F, Yang F, Jiang X, Yu W, Ren X. High-mobility group nucleosome-binding protein 1 is a novel clinical biomarker in non-small cell lung cancer. Tumour Biol 2015;36(12):9405–10.

[7] Mori S, Kita J, Kato M, Shimoda M, Kubota K. Usefulness of a new inflammation-based scoring system for prognostication of patients with hepatocellular carcinoma after hepatectomy. Am J Surg 2015;209(1):187–93.

[8] Jiang H, Wang H, Wang S, Pei Z, Fu Z, Fang C, et al., Expression of ERCC1, TYMS, RRM1, TUBB3, non-muscle myosin II, myoglobin and MyoD1 in lung adenocarcinoma pleural effusions predicts survival in patients receiving platinum-based chemotherapy. Mol Med Rep 2015;11(5):3523–32.

[9] Wu XS, Shi LB, Li ML, Ding Q, Weng H, Wu WG, et al., Evaluation of two inflammation-based prognostic scores in patients with resectable gallbladder carcinoma. Ann Surg Oncol 2014;21(2):449–57.

[10] Suzuki H, Asami K, Hirashima T, Okamoto N, Yamadori T, Tamiya M, et al., Stratification of malignant pleural mesothelioma prognosis using recursive partitioning analysis. Lung 2014;192(1):191–5.

[11] Bodelon C, Polley MY, Kemp TJ, Pesatori AC, McShane LM, Caporaso NE, et al., Circulating levels of immune and inflammatory markers and long versus short survival in early-stage lung cancer. Ann Oncol 2013;24(8):2073–9.

[12] Menschikowski M, Hagelgans A, Schuler U, Froeschke S, Rosner A, Siegert G. Plasma levels of phospholipase A2-IIA in patients with different types of malignancies: Prognosis and association with inflammatory and coagulation biomarkers. Pathol Oncol Res 2013;19(4):839–46.

[13] Son H-J, Park JW, Chang HJ, Kim DY, Kim BC, Kim SY, et al., Preoperative plasma hyperfibrinogenemia is predictive of poor prognosis in patients with nonmetastatic colon cancer. Ann Surg Oncol 2013;20(9):2908–13.

[14] Dehing-Oberije C, Aerts H, Yu S, De Ruysscher D, Menheere P, Hilvo M, et al., Development and validation of a prognostic model using blood biomarker information for prediction of survival of non-small-cell lung cancer patients treated with combined chemotherapy and radiation or radiotherapy alone (NCT00181519, NCT00573040, and NCT0. Int J Radiat Oncol Biol Phys 2011;81(2):360–8.

[15] Alcover J, Filella X, Luqué P, Molina R, Izquierdo L, Augé JM, et al., Prognostic value of IL-6 in localized prostatic cancer. Anticancer Res 2010;30(10):4369–72.

[16] Seppälä M, Tervo S, Pohjola K, Laranne J, Huhtala H, Toppila-Salmi S, et al., The association and prognostic relevance of cancerous inhibitor of protein phosphatase 2A and inflammation in tongue squamous cell carcinoma. APMIS 2015;123(12):1007–15.

[17] Bergenfelz C, Gaber A, Allaoui R, Mehmeti M, Jirström K, Leanderson T, et al., S100A9 expressed in ER(-)PgR(-) breast cancers induces inflammatory cytokines and is associated with an impaired overall survival. Br J Cancer 2015;113(8):1234–43.

[18] Youssef RF, Kapur P, Mosbah A, Abol-Enein H, Ghoneim M, Lotan Y. Role of fibroblast growth factor in squamous cell carcinoma of the bladder: Prognostic biomarker and potential therapeutic target. Urol Oncol 2015;33(3):111.e1–7.

[19] Zong L, Chen P, Wang DX. Death decoy receptor overexpression and increased malignancy risk in colorectal cancer. World J Gastroenterol 2014;20(15):4440–5.

[20] Wouters M, Dijkgraaf E, Kuijjer M, Jordanova E, Hollema H, Welters M, et al., Interleukin-6 receptor and its ligand interleukin-6 are opposite markers for survival and infiltration with mature myeloid cells in ovarian cancer. Oncoimmunology 2014;3(12):e962397.

[21] Dumitru CA, Bankfalvi A, Gu X, Zeidler R, Brandau S, Lang S. AHNAK and Inflammatory Markers Predict Poor Survival in Laryngeal Carcinoma. PLoS One 2013;8(2):e56420.

[22] Huang Y, Zhang J, Cui ZM, Zhao J, Zheng Y. Expression of the CXCl12/CXCR4 and CXCL16/CXCR6 axes in cervical intraepithelial neoplasia and cervical cancer. Chin J Cancer 2013;32(5):289–96.

[23] Morikawa T, Baba Y, Yamauchi M, Kuchiba A, Nosho K, Tanaka N, et al., STAT3 Expression, molecular features, inflammation patterns and prognosis in a database of 724 colorectal cancers. Clin Cancer Res 2011;17(6):617–32.

[24] Hoki Y, Murata M, Hiraku Y, Ma N, Matsumine A, Uchida A, et al., 8-Nitroguanine as a potential biomarker for progression of malignant fibrous histiocytoma, a model of inflammation-related cancer. Oncol Rep 2007;18(5):1165–9.

[25] Hoki Y, Hiraku Y, Ma N, Murata M, Matsumine A, Nagahama M, et al., iNOS-dependent DNA damage in patients with malignant fibrous histiocytoma in relation to prognosis. Cancer Sci 2007;98(2):163–8.

[26] Hilmy M, Campbell R, Bartlett JMS, McNicol A-M, Underwood MA, McMillan DC. The relationship between the systemic inflammatory response, tumour proliferative activity, T-lymphocytic infiltration and COX-2 expression and survival in patients with transitional cell carcinoma of the urinary bladder. Br J Cancer 2006;95(9):1234–8.

[27] Kim B-H, Jung WY, Lee H, Kang Y, Jang Y-J, Hong SW, et al., Lysyl-tRNA synthetase (KRS) expression in gastric carcinoma and tumor-associated inflammation. Ann Surg Oncol 2014;21(6):2020–7.

[28] Belmont L, Rabbe N, Antoine M, Cathelin D, Guignabert C, Kurie J, et al., Expression of TLR9 in tumor-infiltrating mononuclear cells enhances angiogenesis and is associated with a worse survival in lung cancer. Int J Cancer 2014;134(4):765–77.

[29] Fan B, Zhang LH, Jia Y, Zhong XY, Liu YQ, Cheng XJ, et al., Presence of S100A9-positive inflammatory cells in cancer tissues correlates with an early stage cancer and a better prognosis in patients with gastric cancer. BMC Cancer 2012;12:316.

[30] Akaike H, Kono K, Sugai H, Takahashi A, Mimura K, Kawaguchi Y, et al., Expression of high mobility group box chromosomal protein-1 (HMGB-1) in gastric cancer. Anticancer Res 2007;27(1A):449–57.

[31] Hanahan D, Weinberg RA. Hallmarks of cancer: The next generation. Cell 2011;144(5):646–74.

[32] Ferrara N, Hillan KJ, Gerber H-P, Novotny W. Discovery and development of bevacizumab, an anti-VEGF antibody for treating cancer. Nat Rev Drug Discov 2004;3(5):391–400.

[33] Keating GM. Bevacizumab: A review of its use in advanced cancer. Drugs 2014;74(16):1891–925.

[34] Gerber DE. Targeted therapies: A new generation of cancer treatments. Am Fam Physician 2008;77(3):311–9.

[35] Drucker E, Krapfenbauer K. Pitfalls and limitations in translation from biomarker discovery to clinical utility in predictive and personalised medicine. EPMA J 2013;4:7.

[36] Topalian SL, Taube JM, Anders RA, Pardoll DM. Mechanism-driven biomarkers to guide immune checkpoint blockade in cancer therapy. Nat Rev Cancer 2016;16:275–87.

In: Oncology
Editors: A. Furfari et al.

ISBN: 978-1-53610-608-4

Chapter 10

GENETIC BIOMARKERS OF CANCER PROGNOSIS

Bo Angela Wan[1], MPhil, Michael Lam[1], BMSc(C), Anthony Furfari[1], MD(C), Leigha Rowbottom[1], MD(C), Rachel McDonald[1], MD(C), Stephanie Chan[1], BSc(C), Pearl Zaki[1], BSc(C), Carlo DeAngelis[1], PharmD, Azar Azad[2,3,4], PhD, Ronald Chow[1], BMSc(C), Henry Lam[1], MLS, Edward Chow[1], MBBS and George S Charames[3,4,5,*], PhD

[1]Odette Cancer Centre, Sunnybrook Health Sciences Centre, University of Toronto, Toronto, Ontario, Canada
[2]Mount Sinai Services Inc., Toronto, Ontario, Canada
[3]Pathology and Laboratory Medicine, Mount Sinai Hospital, Toronto, Ontario, Canada
[4]Laboratory Medicine and Pathobiology, University of Toronto, Toronto, Ontario, Canada
[5]Lunenfeld-Tanenbaum Research Institute, Sinai Health System, Toronto, Ontario, Canada

---

[*] Correspondence: George S Charames PhD, Pathology and Laboratory Medicine, Mount Sinai Hospital, 600 University Ave, Toronto, Ontario, Canada. E-mail: George.Charames@sinaihealthsystem.ca.

Prognosis after a cancer diagnosis ranges widely depending on the type of cancer, as well as on individual characteristics of the tumour and patient. Historically, most biomarkers used clinically have been biologically based; however, the genetic revolution of the last decade has spurred the investigation and discovery of numerous genetic biomarkers, which relate to aspects of DNA or RNA that are associated with cancer prognosis. In this chapter we summarize the currently available literature reporting on genetic biomarkers and their association with prognosis in cancer patients. A literature search was conducted on Medline and Embase utilizing keywords such as "cancer," "inflammation," "prognosis," and "biomarkers." Articles that reported on genetic biomarkers relating to prognosis were selected for inclusion. Information regarding patient population, cancer type, interventions received, type of genetic marker (i.e., DNA or RNA), and the impact on prognosis was extracted. A total of 18 studies were included in this chapter. Nine studies reported on DNA biomarkers relating to prognosis and the remaining nine studies investigated RNA biomarkers of prognosis. A large amount of literature exists on the association of genetic polymorphisms and gene expression with cancer prognosis. As the number of cancer diagnoses is only expected to increase, genetic biomarkers associated with enhanced or poorer overall survival may have clinical utility and meaning to both oncologists and patients alike.

INTRODUCTION

The prognosis after a cancer diagnosis ranges widely depending on the primary cancer site, and also tumour and patient characteristics. For instance, pancreatic cancer has one of the shortest life expectancies at 2.81 years, while breast cancer has one of the longest, at 20.01 years (1). Variation of prognosis within a single cancer type may be due to multiple factors, including cancer stage at diagnosis, patient lifestyle, and the patient's genetic or physiological dispositions. There is an ongoing effort to examine associations between prognosis and biomarkers, which are certain measurable aspects of a patient's physiology indicative of their state and condition. Historically, biomarkers have largely been biological, such as the presence of certain blood proteins or cell types. However, the genetic revolution of the last decade has spurred the investigation and discovery of numerous genetic biomarkers, which are aspects of DNA or RNA that are associated with cancer prognosis.

There are several types of genetic biomarkers, which can be categorized based on whether they are DNA or RNA molecules (2). DNA biomarkers can be a heritable variation which includes single nucleotide polymorphisms (SNPs), insertion/deletions, variations in short sequence repeats, copy number variations, cytogenetic rearrangements, or epigenetic DNA modifications (2). As populations across the world have varying genetic makeups, genetic variation in race may contribute to differences in survival. Therefore, race is often a controlled factor in assessing associations between SNPs and outcomes. Alternatively, somatic, or non-heritable genetic changes can occur. This can be facilitated by genomic instability, whereby high mutation rates enable the development and malignant transformation of tumours, which may be characterized by their DNA mutation signature biomarkers (3). RNA biomarkers include variations in RNA sequence, levels of mRNA or microRNA (miRNA), or different forms of RNA due to variation in RNA editing processes (2). Both mRNA and miRNA are the transcription products of genes, but while mRNA is translated into a protein, miRNAs are independently functional as short-length RNAs that negatively regulate protein levels by binding to the 3' untranslated region (UTR) of mRNAs and reducing their translation (4).

As differences in DNA may map onto differences in RNA, there is a close relationship between the two classes of genetic biomarkers. For example, DNA variation in the coding region of a gene may result in an mRNA molecule that produces an abnormal protein product, leading to association with both the gene and its mRNA with prognosis. Alternatively, DNA variants at non-coding regions that do not affect protein structure may instead influence the amount or stability of mRNA, and therefore the amount of protein product. A third possibility that applies to SNPs is that the association between an identified genetic variant and prognosis is due to genetic linkage of the candidate SNP with a second unidentified variation, such that they are inherited together. In this case, the identified SNP may not be the functional cause of the associated change in cellular physiology.

Studies on genetic biomarkers that influence patient outcomes provide the basis for the selection of genetic molecular targets for detailed mechanistic and biochemical analysis in order to elucidate pathways of cancer progression, and ultimately provide therapeutic targets or predictors of prognosis. Our chapter aims to provide a summary of genetic biomarkers in current literature that are associated with prognosis in different cancer types.

OUR SEARCH DESIGN

A comprehensive literature search was conducted in Ovid MEDLINE (R) spanning 1946 to June Week 1 2016 and Embase Classic + Embase spanning 1947 to 2016 Week 24. Search terms included 'neoplasm,' 'inflammation,' 'biomarker,' 'allele,' and 'genomics.' The search was limited to English and human subjects.

Study selection

References obtained from the literature search were subjected to title and abstract screening by three authors (AF, AW, ML). References were included if they discussed 1) cancer patients, 2) prognosis, and 3) genetic biomarkers associated with prognosis. References that 1) lacked full text, 2) were review/meta-analysis/opinion/case studies 3) were animal or cell studies, 4) did not include number of patients assessed, or 5) assessed effects of clinical trials were excluded. For studies that met the inclusion and exclusion criteria, full texts were obtained and reviewed by the two authors (AW, ML). If a discrepancy arose, discussion ensued until a consensus was reached. If no consensus could be reached, another author (AF) was consulted and a decision on inclusion was made.

Data extraction

Information from each selected study was extracted, including patient sample size, patient nationality or race, cancer type, intervention, significant DNA or RNA biomarker, significant SNPs, p values, and associated impact on prognosis. The data extracted from each article was reviewed by 2 authors (AW, ML) for accuracy and quality assurance.

OUR FINDINGS

A total of 3,287 original articles were identified using Medline, and 2,193 using Embase. A summary of the screening process can be found in Figure 1. Table 1 summarizes the 18 papers included, comprised of nine studies of DNA biomarkers (SNPs) and nine studies on RNA biomarkers. Sample sizes ranged from 35 to 1,308 patients (5–22). Of the 23 SNPs associated with cancer prognosis, 13 were from eight studies that associated outcomes to individual SNPs; the remaining eight SNPs were from one study that associated prognostic outcomes to a multi-SNP model (7, 10–12, 16, 17, 20–22). There were 52 genes whose mRNA or miRNA levels were associated with prognosis, with two genes found to be significant in more than one study (5, 6, 8, 9, 14, 15, 18, 19, 23).

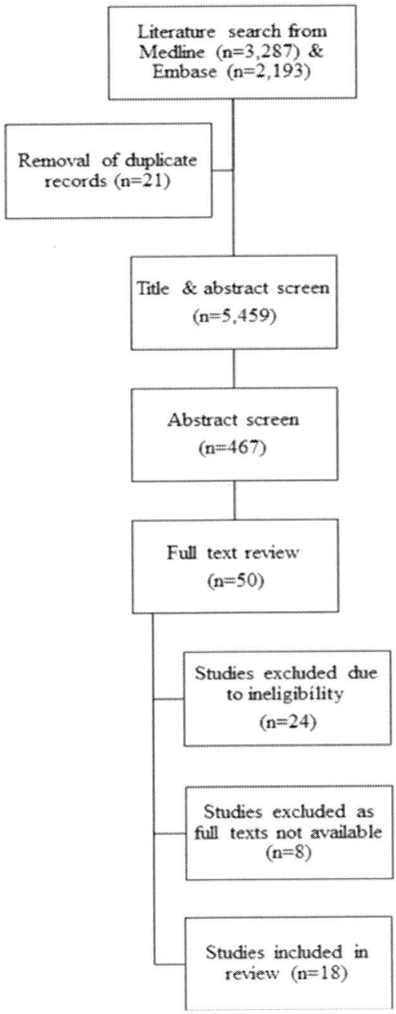

Figure 1. PRISMA flow diagram.

Table 1. Summary of study characteristics

Study Characteristics	No. of studies	
	n	%
Type of genetic biomarker		
DNA (SNPs)	9	50%
mRNA	9	50%
miRNA	1	6%
mRNA and miRNA	1	6%
Cancer type		
Colorectal cancer (CRC)	4	22%
Esophageal cancer	3	17%
Prostate Cancer	3	17%
Hepatocellular carcinoma (HCC)	2	11%
Non-small cell Lung Cancer (NSCLC)	2	11%
Other	4	22%
Interventions		
Chemotherapy	10	56%
Surgery	9	50%
Radiotherapy	5	28%
Other Therapy	2	11%
Multiple	8	44%
Not specified	3	17%
Outcome Assessed		
Overall survival (OS)	17	94%
Cancer specific survival (CSS)	1	6%
Recurrence/disease free survival (DFS)	4	22%
Progression risk/Progression-free survival (PFS)	2	11%
Multiple	4	22%
Total	**18**	

Of the 17 studies that described interventions received by the sample population, the most common treatment was chemotherapy (n=10), followed by surgery (n=9), and radiotherapy (n=5) (5–9, 11–17, 19–22). Eight studies contained multiple types of treatments, and two studies fell under 'other therapy,' which include immunotherapy and hormone therapy (5, 11, 12, 14–17, 22). Across all of the studies included in the review, nine cancer types were investigated, the most common of which was colorectal cancer (CRC) (n=4), esophageal cancer (n=3), and prostate cancer (n=3) (5–22). Overall survival (OS) was the most common outcome assessed (n=17); additional outcomes include recurrence or disease-free survival (n=4) and progression-free survival (n=2) (5–22). Four studies assessed for more than one outcome (5, 8, 9, 12).

Table 2. Genetic biomarkers predictive of prognosis in colorectal, prostate, esophageal, lung, and hepatocellular cancer

Study, year (reference)	Patient race/Country	n	Intervention	Biomarker gene or gene product	Conclusion of polymorphism or RNA expression levels on prognosis
A) Colorectal cancer					
Ting et al. 2013 (21)	Chinese	282	Surgery	*IL10*	rs3021094 GG polymorphism associated with poor OS (p=0.007)
Coghill et al. 2011 (7)	Caucasian	284	Not specified	*IKBKB*	rs11986055 AC/CC polymorphism associated with better CSS (p=0.04) R526Q AG/AA polymorphism associated with worse CSS (p=0.04) and OS (p<0.01)
				PTGS1	rs1213266 AG/AA polymorphism associated with better CSS (p=0.02), rs10306115 AG/AA, rs4836885 AG/GG, L237M CA/AA polymorphisms associated with poor CSS and OS (p≤0.05 for all)
Sharma et al. 2008 (20)	not specified	52	Chemotherapy (Capecitabine)	*IL10*	rs1800872 AA polymorphism associated with worse OS, rs1800896 CC polymorphism associated with better OS (p≤0.05 for both)
Wilkening et al. 2008 (22)	Caucasian	308	Surgery, chemotherapy, radiotherapy	*IL4*	rs2243250 TT/CT polymorphism associated with better OS (p<0.05)
				IL6	rs1800795 CG polymorphism associated with better OS survival (p<0.05)

Table 2. (Continued)

Study, year (reference)	Patient race/Country	n	Intervention	Biomarker gene or gene product	Conclusion of polymorphism or RNA expression levels on prognosis
B) Prostate Cancer					
Dwivedi et al. 2015 (10)	Indian	291	Not specified	*IL10*	-819 CC (promoter) polymorphism associated with higher IL10 protein levels and poor OS (p<0.05)
				IL18	rs187238 GG and -607CC polymorphisms associated with higher IL18 protein levels and better OS (p<0.05 for both)
Olmos et al. 2012 (15)	UK	94	chemotherapy, hormone therapy	*SLC4A1, TFDP1, STOM, HMBS, SNCA, TERF2IP, TMCC2, RIOK3, GABARAPL2*	9-gene model of high mRNA levels of *SLC4A1*, *TFDP1*, *STOM*, *HMBS, SNCA, TERF2IP*, and low mRNA levels of *TMCC2, RIOK3, GABARAPL2* associated with poor OS (p<0.0001)
Ross et al. 2012 (18)	UK	202	Not specified	*ABL2, SEMA4D, ITGAL, C1QA TIMP1, CDKN1A*	6-gene model of high mRNA levels of *ABL2*, *SEMA4D*, *ITGAL* and low mRNA levels of *C1QA, TIMP1, CDKN1A* associated with poor OS (p=0.0067)

Study, year (reference)	Patient race/Country	n	Intervention	Biomarker gene or gene product	Conclusion of polymorphism or RNA expression levels on prognosis
C) Esophageal Cancer					
Zhao et al. 2013 (23)	Chinese	178	surgery	*miR-21, miR-181b, miR-146b, CRY61, CTGF, IL18, VEGF*	High *miR-21, miR-181b* and *miR-146b* in tumour tissue and high *miR-21* in noncancerous tissue are associated with low OS (p=0.0001). High inflammatory risk score consisting of high *CRY61* and *CTGF* and low *IL18* mRNA in tumour tissue and high *VEGF* mRNA in noncancerous tissue associated with poor OS (p=0.0002 training; p=0.005 validation).
Nguyen et al. 2010 (14)	US and Canada	93	surgery, chemotherapy (Cisplatin, 5-fluorouracil), radiotherapy	*IFNG, IL1A, IL18, IL21, IL23, PRG*	6-gene model of high *IFNG, IL1A, IL18, IL21, IL23, PRG* mRNA associated with poor OS (p=0.00026). Association is stronger in Barrett's.
Saadi et al. 2010 (19)	UK	175	unspecified	*TMEPAI, JMY, TSP1, FAP, BCL6*	5-gene model of high *TMEPAI, JMY, TSP1, FAP, BCL6* mRNA associated with poor OS (p=0.022)
D) Non-small cell lung cancer (NSCLC)					
Pu et al. 2014 (16)	Caucasian	1208	Chemotherapy, radiotherapy	*HLA-DOB*	*HLA-DOB* rs2071554 missense variation and *KLRK1* rs2900420

Table 2. (Continued)

Study, year (reference)	Patient race/Country	n	Intervention	Biomarker gene or gene product	Conclusion of polymorphism or RNA expression levels on prognosis
He et al. 2015 (11)	Chinese	1165	Surgery, chemotherapy	*IKBKE, IL1R1, IL6R, NFKB1A, RELB, STAT3*	8-SNP model containing rs10836, rs3732131, rs3732133, rs4072391, rs2273650, rs1053023, rs3744483 and rs28372683, predictive of OS ($p<0.001$)
E) Hepatocellular carcinoma					
Cornella et al. 2015 (8)	US and Europe	58	surgery	*PEAR1, KRTAP5.7, KLRD1, OSBPL8, RPL32, SLC26A11, RGS11 RAPGEF1*	8-gene profile of high mRNA levels of *RAPGEF1, RGS11, OSBPL8, KLRD1, PEAR1, KRTAP5.7,* and decreased *SLC26A11, RPL32* mRNA associated with poor OS and earlier reoccurrence ($p<0.001$)
Dai et al. 2009 (9)	China	110	surgery	*HIF1A, PTGS2, PDGFRA*	High *HIF1A, PDGS2, PDGFRA* mRNA associated with poor OS ($p=0.012$, 0.004, 0.010 respectively) and DFS ($p=0.004$, 0.010, 0.038 respectively)

Table 2A summarizes the significant genetic biomarkers in CRC, corresponding to a total patient sample size of 1,023 (7, 20–22). Eleven significant SNPs from four immune-related genes were associated with survival (7, 20–22). These include five variants in three interleukin genes: *IL10*, which codes for a pleiotropic regulatory cytokine with immunosuppressive and anti-angiogenic properties, *IL6*, coding for a pro-inflammatory cytokine, and *IL4*, coding for a cytokine involved in the helper T cell response (24–26). Four variants in *prostaglandin-endoperoxide synthase 1* (*PTGS1*, or *Cyclooxygenase1*) and two variants in *inhibitor of kappa light polypeptide gene enhancer in B-cells* (*IKBKB*) were also found (7).

Table 2B summarizes the results of three studies on prostate cancer that assessed a total of 587 patients (10, 15, 18). The DNA biomarkers consisted of three SNPs from two cytokine genes, including one *IL10* SNP associated with better OS, and two *IL18* SNPs associated with poor OS (10). Olmos et al. performed mRNA expression arrays on blood to identify a 9-gene signature for predicting survival of metastatic castration-resistant prostate cancer (15). Similarly, Ross et al. identified a prognostic 6-gene signature out of 168 inflammation and prostate-cancer related genes (18).

Three studies on esophageal cancer evaluated 446 patients for mRNA levels that are associated with prognosis (see Table 2C) (14, 19, 23). A study by Zhao et al. found significant RNA and miRNA associated with prognosis. One of these is the microRNA miR-21, a regulator of cell proliferation. High miR-21 levels were associated with poor outcome in the form of advanced disease and reduced survival (23). Zhao et al. and Nguyen et al. both calculated an inflammatory risk score (IRS) to predict survival which involved 6 genes, with *IL18* in common between the two studies (14, 23). However, while Zhao et al. found high *IL18* levels were associated with longer survival, the opposite was found by Nguyen et al; the weight of *IL18* in the IRS was moderate in both. Finally, Saadi et al. performed a gene array assessment and found five genes significantly upregulated in individuals with poor OS, including *thrombospondin 1* (*TSP1*) and *transmembrane prostate androgen induced* (*TMEPAI*) of the *transforming growth factor-beta* (*TGFb*) pathway, as well as the transcription factor *BCL6*, a matrix remodeling protein *junction-mediating and regulatory protein* (*JMY*) and *fibroblast activating protein* (*FAP*) (19).

Two studies, described in Table 2D, examined biomarkers of DNA immune related genes relating to prognosis in a total of 2,373 patients with non-small cell lung cancer (NSCLC) (7, 8). He et al. showed that a group of eight SNPs from six genes in the immune pathway (three genes in the interleukins signaling pathway-*IL1R1, IL6R, STAT3*, and three genes encoding proteins in the nuclear factor kappa B (NFκB) signaling pathway-*NFKB1A, IKBKB, RELB*) were predictive of OS in a Chinese population (11). Pu et al. studied NSCLC in a Caucasian population, and concluded that the missense SNP rs2071554 of the human leukocyte antigen *HLP-DOB* gene was associated with poor OS (OS) (16).

Associations between gene expression and prognosis in hepatocellular carcinoma (HCC) were evaluated in two studies summarized in Table 2E, comprising of 168 patients. Cornella et al. studied 58 HCC patients and found mRNA levels of eight genes were associated with survival and recurrence, including those involved in the PI3K/PTEN and GTP/Ras protein signaling pathways (8). Dai et al. found that high expression levels of *platelet-derived growth factor receptor alpha polypeptide* (*PDGFRA*), *hypoxia-inducible factor 1alpha* (*HIF1A*), and *prostaglandin-endoperoxide synthase 2* (*PTGS2*, or *COX2*) were associated with poor prognosis.

Primary tumours with single studies that found significant prognostic genetic biomarkers are described in Table 3, in order of study size from largest to smallest. Table 3A describes a single study by Reid-Lombardo et al. investigating 1,308 Caucasian patients with pancreatic cancer (17). SNPs in the *insulin-like growth factor 1 receptor* (*IGF1R*) gene, the *mitogen-activated protein kinase 8 interacting protein 1* (*MAPK8IP1*) gene, and the inflammatory pathway *suppressor of cytokine signaling 3* (*SOCS3*) gene were associated with improved survival (17). Another single study consisted of 519 bladder cancer patients and found that the *CC* genotype of the rs1800795 polymorphism in *IL6* is associated with increased risk of progression of non-muscle invasive bladder cancer, yet with improved survival in invasive bladder cancer (see Table 3B) (12). Table 3C describes the study by Asgharzadeh et al. on 244 neuroblastoma patients, who found that high mRNA levels of five inflammatory and tumour cell-related genes are associated with advanced disease and metastasis (5). Finally, Table 3D describes a study by Beiche et al. that investigated rates of recurrence of inflammatory breast cancer (IBC) and non-inflammatory BC and found that mRNA levels of *Sonic Hedgehog* (*SHH*), *N-Myc* (*MYCN*), and *ERG* were of prognostic value (6). The study concluded that relapse was associated with high levels of *SHH* and low levels of *MYCN* and *ERG* (6).

Table 3. Genetic biomarkers predictive of prognosis in pancreatic cancer, bladder cancer, neuroblastoma, and breast cancer

Study, year (reference)	Race/Country	n	Intervention	Biomarker gene or gene product	Conclusion of polymorphism or RNA expression levels on prognosis
A) Pancreatic Cancer					
Reid-Lombardo et al. 2013 (17)	Caucasian	1308	Surgery, chemotherapy, radiotherapy	*IGF1R*	rs1124736 *GG* polymorphism associated with better OS ($p=0.0002$)
				MAPK8IP1	rs3824872 *AC/AA* polymorphism associated with better OS ($p=0.0009$)
				SOCS3	rs8064821 *AC/AA* polymorphism associated with better OS ($p=0.0042$)
B) Bladder cancer					
Leibovici et al. 2005 (12)	Caucasian	519	Surgery, chemotherapy	*IL6*	rs1800795 *CC* polymorphism associated with increased progression risk in non-muscle-invasive bladder cancer (HR, 1.88; 95% CI, 0.80 to 4.11), but improved OS and DFS in invasive bladder cancer (OR 1.77, CI=95%, 1.25-2.51 and HR, 0.39; 95% CI, 0.15 to 1.00, respectively)

Table 3. (Continued)

Study, year (reference)	Race/Country	n	Intervention	Biomarker gene or gene product	Conclusion of polymorphism or RNA expression levels on prognosis
C) Neuroblastoma					
Asgharzadeh et al. 2012 (5)	US and Europe	224	Chemotherapy, radiotherapy, immunotherapy	*CD33, CD16, IL6R, IL10, FCGR3*	High *CD33, CD16, IL6R, IL10, FCGR3* mRNA levels associated with advanced disease and poor PFS ($p<0.001$) and OS ($p=0.003$)
D) Breast Cancer					
Beiche et al. 2004 (6)	Romania	58	chemotherapy (anthracycline)	*MYCN, ERG, SHH*	High *SHH* mRNA levels associated with recurrence ($p=0.045$), low *MYCN* and *ERG* mRNA levels associated with recurrence ($p=0.018$, 0.026 respectively)

DISCUSSION

The present chapter identified 23 SNPs corresponding to 16 genes, and RNA expression levels of 51 genes that are associated with prognosis across a variety of primary cancer types. A group of genetic prognostic factors recurred across multiple studies examining different cancer populations. The products of significant prognostic genetic biomarkers were often part of immune-related pathways, particularly those involved in inflammation such cytokines, chemokines, and NFκB signaling proteins.

One of the most commonly recurring genes associated with cancer prognosis is *IL6*. Moreover, many of the identified significant prognostic genes are part of the IL6 pro-inflammatory signaling pathway. For example, *IL6R* is the gene for the IL6 receptor. Activation of *IL6R* leads to an intracellular signaling cascade resulting in the dimerization and activation of the transcription factor *signal transducer and activator of transcription 3* (*STAT3*) (25). One SNP in the *IL6R* gene, and two SNPs in the *STAT3* gene were identified by He et al. as part of a predictive model for survival in NSCLC (11). *STAT3* facilitates the transcription of many immunity-related genes, including *IL10*, *suppressor of cytokine signaling 3* (*SOCS3*), and *vascular endothelial growth factor* (*VEGF*) and may be an important indicator of prognosis (24). For instance, Reid-Lombardo et al. identified a SNP in *SOCS3* associated with improved survival of pancreatic cancer, and Zhao et al. identified that high *VEGF* mRNA is associated with poor survival in esophageal cancer (17, 23).

Genes in the NFκB pathway were also commonly found to be associated with cancer prognosis. Activation of NFκB involves a host of molecules acting in concert, including dimerization of the products of the *NFKB1A* and *RELB* genes. Proteins such as IKBKB and IKBKE (also known as IKK2 and IKKE) are regulators of this pathway (27). Three genes in the NFκB pathway, *IKBKE, NFKB1A*, and *RELB*, were found to have variants predictive of survival for NSCLC patients (11). Additionally, Coghill et al. identified two SNPs in *IKBKB* associated with improved survival in colorectal cancer (7).

Genes involved in the angiogenesis pathway are also frequently associated with cancer prognosis. Angiogenesis is identified as an important property in the progression of cancer, as it provides circulation to the cancer environment to allow for increased growth (3). Angiogenesis releases tumour cells from the inherent size constraints placed from the limited speed at which nutrients and waste products may be exchanged by diffusion alone (3). Due to its fast growth and high metabolic rate, cancer tumours exist in a hypoxic environment, which triggers activation of the transcription factor known as hypoxia-inducible factor alpha (HIF1A) (28). This leads to the transcription of genes that facilitate angiogenesis and vascularization, among which is *VEGF*, a previously mentioned protein product of IL6 signaling-induced transcription. A second inducer of angiogenesis is *cysteine-rich 61* (*CRY61*). High expression of *HIF1A* and *CRY61* were found to be respectively associated with poor survival of hepatocellular carcinoma and esophageal carcinoma (9, 23).

One of the limitations of this review is the presence of treatments that may influence patient gene expression status as well as outcome. As treatment may influence biomarker levels, the measured outcome may not be an exact reflection of an individual's genetic expression, but in fact due to treatment given. To help control for this, we excluded studies which exclusively investigated patient response to a treatment. However, we included studies in which patients potentially had different treatments from time of biomarker assessment until death, as well as studies in which patients received various treatments. In either case, it is difficult to ascertain the true impact of genetic biomarkers on prognosis without considering the potential confounding effects of prescribed oncologic interventions.

A summary of the genetic biomarkers of cancer prognosis showed that many significant genes are part of common pathways that are implicated in the prognosis of multiple cancer types. Especially prevalent are signaling pathways involving inflammation, immune response, and angiogenesis, which themselves are also interconnected. Since prognostic biomarkers were identified as members of similar pathways, this offers an investigative opportunity to assess potential prognostic value of additional members involved in these pathways. Moreover, since several prognostic genetic biomarkers have been found across different cancer types, this suggests that there are key genes that may underlie the broader pathogenesis of cancer in general. Given the potential of being generally applicable to multiple cancer types, these biomarkers may have added clinical utility.

ACKNOWLEDGMENT

We thank the generous support of Joey and Mary Furfari Cancer Research Fund.

REFERENCES

[1] Chu PC, Wang JD, Hwang JS, Chang YY. Estimation of life expectancy and the expected years of life lost in patients with major cancers: Extrapolation of survival curves under high-censored rates. Value Heal 2008;11(7):1102–9.

[2] Novelli G, Ciccacci C, Borgiani P, Amati MP, Abadie E. Genetic tests and genomic biomarkers: Regulation, qualification and validation. Clin Cases Miner Bone Metab 2008;5(2):149–54.

[3] Hanahan D, Weinberg RA. Hallmarks of cancer: The next generation. Cell 2011;144(5):646–74.

[4] Hayes J, Peruzzi PP, Lawler S. MicroRNAs in cancer: Biomarkers, functions and therapy. Trends Mol Med 2014;20(8):460–9.

[5] Asgharzadeh S, Salo JA, Ji L, Oberthuer A, Fischer M, Berthold F, et al. Clinical significance of tumor-associated inflammatory cells in metastatic neuroblastoma. J Clin Oncol 2012;30(28):3525–32.

[6] Beiche I, Lerebours F, Tozlu S, Espie M, Marty M, Lidereau R. Molecular profiling of inflammatory breast cancer : identification of a poor-prognosis gene expression signature. Clin Cancer Res 2004;10:6789–95.

[7] Coghill AE, Newcomb PA, Poole EM, Hutter CM, Makar W, Duggan D, et al. Genetic variation in inflammatory pathways is related to colorectal cancer survival. Clin Cancer Res 2011;17(22):7139–47.

[8] Cornella H, Alsinet C, Sayols S, Zhang Z, Hao K, Cabellos L, et al. Unique Genomic profile of fibrolamellar hepatocellular carcinoma. Gastroenterology 2015;148(4):806–18.

[9] Dai CX, Gao Q, Qiu SJ, Ju MJ, Cai MY, Xu YF, et al. Hypoxia-inducible factor-1 alpha, in association with inflammation, angiogenesis and MYC, is a critical prognostic factor in patients with HCC after surgery. BMC Cancer 2009;9:418.

[10] Dwivedi S, Goel A, Mandhani A, Khattri S, Sharma P, Misra S, et al. Functional genetic variability at promoters of pro-(IL-18) and anti-(IL-10) inflammatory affects their mRNA expression and survival in prostate carcinoma patients: Five year follow-up study. Prostate 2015;75(15):1737–46.

[11] He F, Yang R, Li X, Ye C, He B, Lin T, et al. Single nucleotide polymorphisms of the NF-kappaB and STAT3 signaling pathway genes predict lung cancer prognosis in a Chinese Han population. Cancer Genet 2015;208(6):310–8.

[12] Leibovici D, Grossman HB, Dinney CP, Millikan RE, Lerner S, Wang Y, et al. Polymorphisms in inflammation genes and bladder cancer: From initiation to recurrence, progression, and survival. J Clin Oncol 2005;23(24):5746–56.

[13] Na GH, Kim DG, Han JH, Kim EY, Lee SH, Hong TH, et al. Inflammatory markers as selection criteria of hepatocellular carcinoma in living-donor liver transplantation. World J Gastroenterol 2014;20(21):6594–601.

[14] Nguyen GH, Schetter AJ, Chou DB, Bowman ED, Zhao R, Hawkes JE, et al. Inflammatory and microRNA gene expression as prognostic classifier of Barrett's-associated esophageal adenocarcinoma. Clin Cancer Res 2010;16(23):5824–34.

[15] Olmos D, Brewer D, Clark J, Danila DC, Parker C, Fleisher M, et al. Prognostic value of blood mRNA expression signatures in castration-resistant prostate cancer: a prospective, two-stage study. Lancet Oncol 2012;13(11):1114–24.

[16] Pu X, Hildebrandt MA, Lu C, Roth JA, Stewart DJ, Zhao J, et al. Inflammation-related genetic variations and survival for advanced non-small cell lung cancer receiving first-line chemotherapy. Clin Pharmacol Ther 2014;96(3):360–9.

[17] Reid-Lombardo KM, Fridley BL, Bamlet WR, Cunningham JM, Sarr MG, Petersen GM. Survival is associated with genetic variation in inflammatory pathway genes among patients with resected and unresected pancreatic cancer. Ann Surg 2013;257(6):1096-102.

[18] Ross RW, Galsky MD, Scher HI, Magidson J, Magidson K, Lee GSM, et al. A whole-blood RNA transcript-based prognostic model in men with castration-resistant prostate cancer: A prospective study. Lancet Oncol 2012;13(11):1105–13.

[19] Saadi A, Shannon NB, Lao-Sirieix P, O'Donovan M, Walker E, Clemons NJ, et al. Stromal genes discriminate preinvasive from invasive disease, predict outcome, and highlight inflammatory pathways in digestive cancers. Proc Natl Acad Sci U S A. 2010;107(5):2177–82.

[20] Sharma R, Zucknick M, London R, Kacevska M, Liddle C, Clarke SJ. Systemic inflammatory response predicts prognosis in patients with advanced-stage colorectal cancer. Clin Colorectal Cancer 2008;7(5):331–7.

[21] Ting WC, Chen LM, Huang LC, Hour MJ, Lan YH, Lee HZ, et al. Impact of interleukin-10 gene polymorphisms on survival in patients with colorectal cancer. J Korean Med Sci 2013;28(9):1302–6.

[22] Wilkening S, Tavelin B, Canzian F, Enquist K, Palmqvist R, Altieri A, et al. Interleukin promoter polymorphisms and prognosis in colorectal cancer. Carcinogenesis 2008;29(6):1202–6.

[23] Zhao Y, Schetter AJ, Yang GB, Nguyen G, Mathé EA, Li P, et al. microRNA and inflammatory gene expression as prognostic marker for overall survival in esophageal squamous cell carcinoma. Int J Cancer 2013;132(12):2901–9.

[24] Mannino MH, Zhu Z, Xiao H, Bai Q, Wakefield MR, Fang Y. The paradoxical role of IL-10 in immunity and cancer. Cancer Lett 2015;367(2):103–7.

[25] Mauer J, Denson JL, Brüning JC. Versatile functions for IL-6 in metabolism and cancer. Trends Immunol 2015;36(2):92–101.

[26] Brown MA. IL-4 production by T cells: you need a little to get a lot. J Immunol 2008;181(5):2941–2.

[27] Mitchell S, Vargas J, Hoffmann A. Signaling via the NFkB system. Wiley Interdiscip Rev Syst Biol Med 2016;8(3):227–41.

[28] Royds JA, Dower SK, Qwarnstrom EE, Lewis CE. Response of tumour cells to hypoxia: role of p53 and NFkB. Mol Pathol 1998;51(2):55–61.

SECTION TWO: FUTURE ASPECTS

In: Oncology
Editors: A. Furfari et al.

ISBN: 978-1-53610-608-4

Chapter 11

THE PROMISING FUTURE

Stephanie Chan[1], BSc(C), Bo Angela Wan[1], MPhil, Pearl Zaki[1], BSc(C), Adam Bobrowski[1], BSc, Matthew Choi[1], BSc, Carlo DeAngelis[1], PharmD, Azar Azad[2,3,4], PhD, Rachel McDonald[1], MD(C), Leigha Rowbottom[1], MD(C), Anthony Furfari[1], MD(C), Ronald Chow[1], BMSc(C), Edward Chow[1], MBBS and George S Charames[3,4,5,*], PhD

[1]Odette Cancer Centre, Sunnybrook Health Sciences Centre, University of Toronto, Toronto, Ontario, Canada
[2]Mount Sinai Services Inc., Toronto, Ontario, Canada
[3]Pathology and Laboratory Medicine, Mount Sinai Hospital, Toronto, Ontario, Canada
[4]Laboratory Medicine and Pathobiology, University of Toronto, Toronto, Ontario, Canada
[5]Lunenfeld-Tanenbaum Research Institute, Sinai Health System, Toronto, Ontario, Canada

* Correspondence: George S Charames PhD, FACMG, Pathology and Laboratory Medicine, Mount Sinai Hospital, Suite 6-500, 600 University Ave, Toronto, Ontario, M5G 1X5, Canada. E-mail: George.Charames@sinaihealthsystem.ca.

Introduction

During the last decade, countless studies have explored the role of biomarkers in oncology. Biomarkers have provided researchers with clues to understanding the cellular mechanisms behind important biological processes that allow tumours to develop and grow. The potential clinical applications biomarkers have for personalized medicine is endless as long as research in this developing field continues. Biomarkers have already been used to improve treatment in non-small cell lung cancer (NSCLC) patients and breast cancer patients with the development of targeted therapy for patients with tumours positive for the amplification of growth-promoting genes such as *HER2* (1). In the future, biomarkers could be used as prognostic indicators, or even predict patients' response to treatment or analgesics, ultimately arming physicians with more information and tools to improve and personalize each patient's care. This special issue has compiled and summarized the currently available literature regarding biomarkers in oncology, specifically in their ability to predict symptom burden and quality of life (QOL), radiation response and radiation-induced toxicities, drug metabolism, and prognosis.

Pain

Pain is one of the most common symptoms experienced by cancer patients and if improperly managed, can severely impact their QOL. Individualized pain treatment, especially those involving morphine and other opioids, is limited by differences in inter-individual responses to side effects and analgesic effectiveness. The utilization of biological markers can potentially advance the standard of care and best practices leading to the determination of appropriate dosing protocols and improved health outcomes. Our review of biomarkers predicting the efficacy of analgesics and individual patients' susceptibility to pain showed that there are few instances where pain can be determined by a single gene, as many studies analyzing specific single nucleotide polymorphisms (SNPs) have inconsistent conclusions with regards to establishing genetic profiles conferring different pain management outcomes. Future studies must look at the interactions and relationships of multiple genes and assess combined predictive scores involving more than one SNP in order to attain greater predictive power.

The disruption of tissue homeostasis and repair common to both inflammation and cancer has raised the possibility that non-steroidal anti-inflammatory drugs (NSAID) based compounds may be used in the prophylactic treatment of cancer and prevention of recurrence (2). Studies of colorectal, lung and prostate cancer patients have failed to support this hypothesis (3–9). However, use of NSAIDs has been correlated with decreased breast cancer occurrence (10). Due to the inconsistent nature of the findings on this topic, further investigations and pharmacogenetic inquiry is necessary to characterize the use of NSAIDs in the prophylaxis of cancer.

RADIATION

A total of eleven studies assessed the effects of radiation, four of which focused on biomarkers predicting radiation response (11–14), and seven of which focused on radiation toxicities (15–21). Results from studies on radiation response found that several inflammatory cytokine plasma levels (12, 13), an SNP of the gene *RNASEL* (11), and changes in miRNA profiles (14) were significantly correlated with treatment outcomes. Results from studies on radiation toxicities produced results of varying significance, although multiple studies have focused on the role of the pro-inflammatory C-reactive protein (CRP) (15, 20). Although plasma CRP levels were only found to have borderline significant association with radiation-induced toxicities, association studies of genetic variation of several inflammation-related genes show clinical value in its potential to stratify the patient population, especially when used as part of a cumulative polygene risk score to predict toxicity risk (16–19, 21). Risk profiles would better inform patients and physicians about expected radiation outcomes and radiation-induced toxicities, which not only aids in clinical preparation, but also in the decision-making process regarding goals of care on whether to forgo treatment in the event of high toxicity risk in favour of palliative medicine. To achieve this, more data is required on the efficacy of biomarkers in predicting the responses and effects of radiotherapy.

SYMPTOM BURDEN

Symptom burden is a very important endpoint in biomarker studies as cancer patients often experience multiple symptoms simultaneously, drastically decreasing their QOL. High serum levels of CRP were associated with greater symptom burden across multiple studies of cancer patients with different primaries (22–30). Inflammatory cytokines were also correlated with symptom burden and QOL, suggesting the etiology of these symptom clusters is closely related to inflammation pathways. Research and development of genetic and protein biomarkers as well as further investigations of the mechanisms relating inflammation and cancer-related symptoms may allow physicians to improve personalized medicine to treat all symptoms.

PROGNOSIS

Numerous genetic biomarkers including SNPs and miRNAs have been identified as having potential to stratify patients into prognostic and susceptible groups. Genetic, protein-based, and cell-based biomarkers are frequently implicated in inflammation pathways, the activation of which is associated with adverse outcomes. The most commonly investigated prognostic cancer biomarkers are also indicators of systemic inflammation. Common inflammation scores include the neutrophil-lymphocyte ratio (NLR) (31–41), and the Glasgow Prognostic Score (GPS) which is based on plasma levels of CRP and albumin (31, 33, 34, 38–40, 42–50). These have been validated in a number of cancer types, implicating broad significance. Additional protein and cell based biomarkers include compounds found in tissues and bodily fluids, and are often implicated in inflammation, immune, DNA damage, and tissue-remodeling pathways.

HOPE

With new technology making high-throughput genomic sequencing not only possible, but extremely efficient, research in biomarkers and their applications in oncology will continue to grow and expand. As we investigate deeper into the mechanisms of cancer cells and identify more and more biomarkers with clinical correlations, we open up endless possibilities to personalize and improve patient care which is sorely needed in this population.

ACKNOWLEDGMENTS

We thank the generous support of Joey and Mary Furfari Cancer Research Fund.

REFERENCES

[1] Stopeck AT, Brown-Glaberman U, Wong HY, Park BH, Barnato SE, Gradishar WJ, et al. The role of targeted therapy and biomarkers in breast cancer treatment. Clin Exp Metastasis 2012;29(7):807–19.

[2] Hanahan D, Weinberg RA. Hallmarks of cancer: the next generation. Cell 2011;144(5):646–74.

[3] Barry EL, Sansbury LB, Grau M V, Ali IU, Tsang S, Munroe DJ, et al. Cyclooxygenase-2 polymorphisms, aspirin treatment, and risk for colorectal adenoma recurrence--data from a randomized clinical trial. Cancer Epidemiol Biomarkers Prev 2009;18(10):2726–33.

[4] Habermann N, Brown E, Toth R, Scherer D, Buck K, Cheng TYD, et al. Abstract 2186: Inflammation- and angiogenesis- related genes: Interaction with NSAID use, and serum inflammatory markers on colorectal cancer risk within the Women's Health Initiative. Cancer Res 2014;74(19 Supplement) 2186.

[5] Kopp TI, Friis S, Christensen J, Tjønneland A, Vogel U. Polymorphisms in genes related to inflammation, NSAID use, and the risk of prostate cancer among Danish men. Cancer Genet 2013;206(7–8):266–78.

[6] Slattery ML, Wolff RK, Herrick JS, Caan BJ, Potter JD. IL6 genotypes and colon and rectal cancer. Cancer Causes Control 2007;18(10):1095–105.

[7] Kraus S, Hummler S, Toriola AT, Poole EM, Scherer D, Kotzmann J, et al. Impact of genetic polymorphisms on adenoma recurrence and toxicity in a COX2 inhibitor (celecoxib) trial: results from a pilot study. Pharmacogenet Genomics 2013;23(8):428–37.

[8] Vogel U, Christensen J, Wallin H, Friis S, Nexø BA, Raaschou-Nielsen O, et al. Polymorphisms in genes involved in the inflammatory response and interaction with NSAID use or smoking in relation to lung cancer risk in a prospective study. Mutat Res 2008;639(1–2):89–100.

[9] Vogel U, Christensen J, Dybdahl M, Friis S, Hansen RD, Wallin H, et al. Prospective study of interaction between alcohol, NSAID use and polymorphisms in genes involved in the inflammatory response in relation to risk of colorectal cancer. Mutat Res 2007;624(1–2):88–100.

[10] Menegaux F, Truong T, Ajmia F, Anger A, Mulot C, Laurent-Puig P, et al. Abstract 141: Nonsteroidal anti-inflammatory drugs (NSAIDs), cyclooxygenase-2 polymorphisms, and breast cancer risk. Cancer Res 2014;73(8 Supplement) 141.

[11] Schoenfeld JD, Margalit DN, Kasperzyk JL, Shui IM, Rider JR, Epstein MM, et al. A single nucleotide polymorphism in inflammatory gene RNASEL predicts outcome after radiation therapy for localized prostate cancer. Clin Cancer Res 2013;19(6):1612–9.

[12] Barrett K, Xu J, Kim S, Boyd K, Cheng D, Chen Z, et al. Genetic polymorphisms associated with toxicity are associated with overall survival following curative radiation for non-small cell lung cancer. Int J Radiat Oncol 2014;90(1):S811.

[13] Bortolin M, Trovò M, Furlan C, Bidoli E, Basaglia G, Avanzo M, et al. Multiplexed plasma cytokine chemokine and growth factor profiling in early-stage non-small cell lung cancer patients undergoing stereotactic body radiation therapy. Int J Radiat Oncol 2014;90(1):S811–2.

[14] Yu Q, Li B, Li P, Shi Z, Vaughn A, Zhu L, et al. Plasma microRNAs to predict the response of radiotherapy in esophageal squamous cell carcinoma patients. Am J Transl Res 2015;7(10):2060–71.

[15] Rodriguez-Gil JL, Takita C, Wright J, Reis IM, Zhao W, Lally BE, et al. Inflammatory biomarker c-reactive protein and radiotherapy-induced early adverse skin reactions in patients with breast cancer. Cancer Epidemiol Prev Biomarkers 2014;23(9):1873-83.

[16] Pu X, Wang L, Chang JY, Hildebrandt MAT, Ye Y, Lu C, et al. Inflammation-related genetic variants predict toxicity following definitive radiotherapy for lung cancer. Clin Pharmacol Ther 2014;96(5):609–15.

[17] Hildebrandt MAT, Komaki R, Liao Z, Gu J, Chang JY, Ye Y, et al. Genetic variants in inflammation-related genes are associated with radiation-induced toxicity following treatment for non-small cell lung cancer. PLoS One 2010;5(8):e12402.

[18] Ho AY, Fan G, Atencio DP, Green S, Formenti SC, Haffty BG, et al. Possession of ATM sequence variants as predictor for late normal tissue responses in breast cancer patients treated with radiotherapy. Int J Radiat Oncol Biol Phys 2007;69(3):677–84.

[19] Guerra JLL, Gomez D, Wei Q, Liu Z, Wang LE, Yuan X, et al. Association between single nucleotide polymorphisms of the transforming growth factor β1 gene and the risk of severe radiation esophagitis in patients with lung cancer. Radiother Oncol 2012;105(3):299–304.

[20] Mohammed FF, Poon I, Zhang L, Elliott L, Hodson ID, Sagar SM, et al. Acute-phase response reactants as objective biomarkers of radiation-induced mucositis in head and neck cancer. Head Neck 2012;34(7):985–93.

[21] Yuan X, Liao Z, Liu Z, Wang LE, Tucker SL, Mao L, et al. Single nucleotide polymorphism at rs1982073:T869C of the TGFbeta 1 gene is associated with the risk of radiation pneumonitis in patients with non-small-cell lung cancer treated with definitive radiotherapy. J Clin Oncol 2009;27(20):3370–8.

[22] Bower JE, Ganz PA, Tao ML, Hu W, Belin TR, Sepah S, et al. Inflammatory biomarkers and fatigue during radiation therapy for breast and prostate cancer. Clin Cancer Res 2009;15(17):5534–40.

[23] Pertl MM, Hevey D, Boyle NT, Hughes MM, Collier S, O'Dwyer AM, et al. C-reactive protein predicts fatigue independently of depression in breast cancer patients prior to chemotherapy. Brain Behav Immun 2013;34:108–19.

[24] Reinertsen K V, Grenaker Alnæs GI, Landmark-Høyvik H, Loge JH, Wist E, Kristensen VN, et al. Fatigued breast cancer survivors and gene polymorphisms in the inflammatory pathway. Brain Behav Immun 2011;25(7):1376–83.

[25] Penafuerte CA, Gagnon B, Sirois J, Murphy J, MacDonald N, Tremblay ML. Identification of neutrophil-derived proteases and angiotensin II as biomarkers of cancer cachexia. Br J Cancer 2016;114(6):680–7.

[26] Archer JA, Hutchison IL, Dorudi S, Stansfeld SA, Korszun A. Interrelationship of depression, stress and inflammation in cancer patients: a preliminary study. J Affect Disord 2012;143(1–3):39–46.

[27] Imayama I, Alfano CM, Neuhouser ML, George SM, Wilder Smith A, Baumgartner RN, et al. Weight, inflammation, cancer-related symptoms and health related quality of life among breast cancer survivors. Breast Cancer Res Treat 2013;140(1):159–76.

[28] Kao SC, Vardy J, Harvie R, Chatfield M, van Zandwijk N, Clarke S, et al. Health-related quality of life and inflammatory markers in malignant pleural mesothelioma. Support Care Cancer 2013;21(3):697–705.

[29] Burney BO, Hayes TG, Smiechowska J, Cardwell G, Papusha V, Bhargava P, et al. Low testosterone levels and increased inflammatory markers in patients with cancer and relationship with cachexia. J Clin Endocrinol Metab 2012;97(5):E700-9.

[30] Low CA, Bovbjerg DH, Jenkins FJ, Ahrendt SA, Choudry HA, Holtzman MP, et al. Preoperative inflammatory biomarkers and neurovegetative symptoms in peritoneal carcinomatosis patients. Brain Behav Immun 2014;42:65–8.

[31] Aurello P, Tierno SM, Berardi G, Tomassini F, Magistri P, D'Angelo F, et al. Value of preoperative inflammation-based prognostic scores in predicting overall survival and disease-free survival in patients with gastric cancer. Ann Surg Oncol 2014;21(6):1998–2004.

[32] Deng Q, He B, Liu X, Yue J, Ying H, Pan Y, et al. Prognostic value of pre-operative inflammatory response biomarkers in gastric cancer patients and the construction of a predictive model. J Transl Med 2015;13:66.

[33] Kinoshita A, Onoda H, Imai N, Iwaku A, Oishi M, Tanaka K, et al. The C-reactive protein/albumin ratio, a novel inflammation-based prognostic score, predicts outcomes in patients with hepatocellular carcinoma. Ann Surg Oncol 2014;22(3):803–10.

[34] Martin HL, Ohara K, Kiberu A, Van Hagen T, Davidson A, Khattak MA. Prognostic value of systemic inflammation-based markers in advanced pancreatic cancer. Intern Med J 2014;44(7):676–82.

[35] Na GH, Kim DG, Han JH, Kim EY, Lee SH, Hong TH, et al. Inflammatory markers as selection criteria of hepatocellular carcinoma in living-donor liver transplantation. World J Gastroenterol 2014;20(21):6594–601.

[36] Qi Q, Lee IK, Lee YS, Sung NY, Kwon TS. Clinical implications of systemic inflammatory response markers as independent prognostic factors in colorectal cancer patients. Pancreatology 2015;15(2):145–50.

[37] Szkandera J, Stotz M, Eisner F, Absenger G, Stojakovic T, Samonigg H, et al. External validation of the derived neutrophil to lymphocyte ratio as a prognostic marker on a large cohort of pancreatic cancer patients. PLoS One 2013;8(11): e78225.

[38] Tan CSY, Read JA, Phan VH, Beale PJ, Peat JK, Clarke SJ. The relationship between nutritional status, inflammatory markers and survival in patients with advanced cancer: a prospective cohort study. Support Care Cancer 2015;23(2):385–91.

[39] Wang DS, Ren C, Qiu MZ, Luo HY, Wang ZQ, Zhang DS, et al. Comparison of the prognostic value of various preoperative inflammation-based factors in patients with stage III gastric cancer. Tumor Biol 2012;33(3):749–56.

[40] Xu XL, Yu HQ, Hu W, Song Q, Mao WM. A novel inflammation-based prognostic score, the c-reactive protein/albumin ratio predicts the prognosis of patients with operable esophageal Squamous cell carcinoma. PLoS One 2015;10(9): e0138657.

[41] Zhang W, Liu K, Hu G, Liang W. Preoperative platelet/lymphocyte ratio is a superior prognostic factor compared to other systemic inflammatory response markers in ovarian cancer patients. Tumor Biol 2015;36(11):8831–7.

[42] Jiang X, Hiki N, Nunobe S, Kumagai K, Kubota T, Aikou S, et al. Prognostic importance of the inflammation-based Glasgow prognostic score in patients with gastric cancer. Br J Cancer 2012;107(2):275–9.

[43] Kinoshita A, Onoda H, Imai N, Iwaku A, Oishi M, Tanaka K, et al. The Glasgow Prognostic Score, an inflammation based prognostic score, predicts survival in patients with hepatocellular carcinoma. BMC Cancer 2013;13(1):52.

[44] Laird BJ, Kaasa S, McMillan DC, Fallon MT, Hjermstad MJ, Fayers P, et al. Prognostic factors in patients with advanced cancer: a comparison of clinicopathological factors and the development of an inflammation-based prognostic system. Clin Cancer Res 2013;19(19):5456–64.

[45] McMillan DC, Crozier JEM, Canna K, Angerson WJ, McArdle CS. Evaluation of an inflammation-based prognostic score (GPS) in patients undergoing resection for colon and rectal cancer. Int J Colorectal Dis 2007;22(8):881–6.

[46] Mohri Y, Tanaka K, Ohi M, Toiyama Y, Yasuda H, Inoue Y, et al. Inflammation-based prognostic score as a predictor of postoperative gastric cancer recurrence. Anticancer Res 2012;32(10):4581–4.

[47] Nakamura M, Iwahashi M, Nakamori M, Ojima T, Katsuda M, Iida T, et al. A new prognostic score for the survival of patients with esophageal squamous cell carcinoma. Surg Today 2014;44(5):875–83.

[48] Oshiro Y, Sasaki R, Fukunaga K, Kondo T, Oda T, Takahashi H, et al. Inflammation-based prognostic score is a useful predictor of postoperative outcome in patients with extrahepatic cholangiocarcinoma. J Hepatobiliary Pancreat Sci 2013;20(3):389–95.

[49] Partridge M, Fallon M, Bray C, McMillan D, Brown D, Laird B. Prognostication in advanced cancer: A study examining an inflammation-based score. J Pain Symptom Manage 2012;44(2):161–7.

[50] Shibutani M, Maeda K, Nagahara H, Ohtani H, Iseki Y, Ikeya T, et al. The prognostic significance of a postoperative systemic inflammatory response in patients with colorectal cancer. World J Surg Oncol 2015;13(1):194.

SECTION THREE: ACKNOWLEDGMENTS

In: Oncology
Editors: A. Furfari et al.
ISBN: 978-1-53610-608-4

Chapter 12

About the Editors

Anthony Furfari, MD(C) is a research assistant in the Rapid Response Radiotherapy Program at the Odette Cancer Center, Sunnybrook Health Sciences, Toronto, Canada and also a medical student at the Royal College of Surgeons Ireland, Bahrain.
E-mail: Anthony.Furfari@yahoo.ca

George S Charames, PhD, DABMG, FACMG, is the Director of the Advanced Molecular Diagnostics Laboratory in Mount Sinai Hospital's Department of Pathology and Lab Medicine, Assistant Professor in the Department of Laboratory Medicine and Pathobiology at the University of Toronto, and an Associate Member and Scientist in the Lunenfeld-Tanenbaum Research Institute, Sinai Health System. Dr. Charames received his MSc and PhD from the Department of Lab Medicine and Pathobiology, University of Toronto, before completing his Clinical Molecular Genetics Postdoctoral Fellowship at the Johns Hopkins School of Medicine in Baltimore, United States. He is a board certified Diplomate of the American Board of Medical Genetics and Genomics (ABMGG), and a Fellow of the American College of Medical Genetics and Genomics (ACMG).
E-mail: George.Charames@sinaihealthsystem.ca

Rachel McDonald, MD(C) is a clinical research assistant in the Rapid Response Radiotherapy Program at the Odette Cancer Center, Sunnybrook Health Sciences, Toronto, Canada and also a medical student at the University of Toronto, Canada. She is the receipient of the national Co-op Student of the Year Award from the Canadian Association for Co-operative Education

(CAFCE) and also the provincial Co-op Student of the Year Award from Education at Work Ontario (EWO) in 2016.
E-mail: Rachel.McDonald@sunnybrook.ca

Leigha Rowbottom, MD(C) is a clinical research assistant in the Rapid Response Radiotherapy Program at the Odette Cancer Center, Sunnybrook Health Sciences, Toronto, Canada. She is also a medical student at the University of Toronto, Canada. She was the recipient of the Jocelyn Cowan Proudly She Marched Scholarship from the University of Waterloo in 2015 for academic excellence and extracurricular involvement.
E-mail: Leigha.Rowbottom@sunnybrook.ca

Azar Azad, MLD, PhD is a lecturer at the University of Toronto, a scientific staff at Mount Sinai Hospital in Toronto, the CEO of Mount Sinai Services Inc. and Co-Director of Banting and Best Diabetes Center Core Laboratories. Her focus has been on angiogenesis, apoptosis, cancer, cardiovascular and diabetes. Her team has been involved in multiple FDA submissions for medical devices, platforms, multiplex assays and successfully adopted new innovative technologies to healthcare applications.
E-mail: Azar.Azad2@sinaihealthsystem.ca

Stephanie Chan, BSc(C) is a clinical research assistant in the Rapid Response Radiotherapy Program at the Odette Cancer Centre, Sunnybrook Health Sciences Centre, Toronto, Canada and also a biochemistry student at the University of Waterloo, Canada. She was the recipient of the J Frank Brookfield Scholarship for excellence in biology and Don E Irish Scholarship in Science from the University of Waterloo in 2016.
E-mail: Stephanie.Chan@sunnybrook.ca

Bo Angela Wan, BA, MPhil is a clinical research assistant in the Rapid Response Radiotherapy Program at the Odette Cancer Centre, Sunnybrook Health Sciences Centre, Toronto, Canada and completed her education at the University of Cambridge in the United Kingdom under the Blyth Cambridge Commonwealth Trust Scholarship. She previously conducted research on the neurobiology of fruit fly olfaction and her current interests lie in cancer biomarkers and palliative medicine.
E-mail: Bo.Wan@cantab.net

Ronald Chow, BMSc(C) is a clinical research assistant in the Rapid Response Radiotherapy Program at the Odette Cancer Centre, Sunnybrook Health Sciences Centre, Toronto, Canada. He was a recipient of the Lieutenant Governor's Community Volunteer Award, Chancellor's Scholarship from Queen's University for superior academic ability and leadership skills, and George Eastman Young Leaders Award from the University of Rochester for astounding leadership and extensive extracurricular activities.
E-mail: Ronald.Chow@sunnybrook.ca

Carlo DeAngelis, PharmD is the Clinical Pharmacy Coordinator for Oncology at the Odette Cancer Centre, Sunnybrook Health Sciences Centre, Toronto, Canada. He is also the Clinician Scientist and Assistant Professor of the Leslie Dan Faculty of Pharmacy at the University of Toronto, Canada. He was named Pharmacist of the Year by the Ontario Pharmacists Association in 2016.
E-mail: Carlo.DeAngelis@sunnybrook.ca

Pearl Zaki, BSc(C) is a clinical research assistant in the Rapid Response Radiotherapy Program at the Odette Cancer Centre, Sunnybrook Health Sciences Centre, Toronto, Canada and also a health studies student at the University of Waterloo, Canada. She was the recipient of the University of Waterloo President's Scholarship for academic excellence in 2014.
E-mail: Pearl.Zaki@sunnybrook.ca

Edward Chow, MBBS, MSc, PhD, FRCPC is Professor of Radiation Oncology at the University of Toronto in Canada. He is the chair of the Rapid Response Radiotherapy Program and Bone Metastases Site Group at the Odette Cancer Centre, Sunnybrook Health Sciences Centre, Toronto, Canada and also a senior scientist at the Sunnybrook Research Institute. He has published in the area of palliative radiotherapy and end of life care issues.
E-mail: Edward.Chow@sunnybrook.ca

Joav Merrick, MD, MMedSci, DMSc, born and educated in Denmark is professor of pediatrics, child health and human development, Division of Pediatrics, Hadassah Hebrew University Medical Center, Mt Scopus Campus, Jerusalem, Israel and Kentucky Children's Hospital, University of Kentucky, Lexington, Kentucky United States and professor of public health at the Center for Healthy Development, School of Public Health, Georgia State University,

Atlanta, United States, the medical director of the Health Services, Division for Intellectual and Developmental Disabilities, Ministry of Social Affairs and Social Services, Jerusalem, the founder and director of the National Institute of Child Health and Human Development in Israel. Numerous publications in the field of pediatrics, child health and human development, rehabilitation, intellectual disability, disability, health, welfare, abuse, advocacy, quality of life and prevention. Received the Peter Sabroe Child Award for outstanding work on behalf of Danish Children in 1985 and the International LEGO-Prize ("The Children's Nobel Prize") for an extraordinary contribution towards improvement in child welfare and well-being in 1987.
E-mail: jmerrick@zahav.net.il

In: Oncology
Editors: A. Furfari et al.
ISBN: 978-1-53610-608-4

Chapter 13

ABOUT THE RAPID RESPONSE RADIOTHERAPY PROGRAM AT THE ODETTE CANCER CENTRE, SUNNYBROOK HEALTH SCIENCES CENTRE, TORONTO, CANADA

The Odette Cancer Centre, the comprehensive cancer program of Sunnybrook Health Sciences Centre is a leading regional cancer centre in Toronto, Ontario, Canada. It is the sixth largest cancer centre in North America in terms of number of new cancer patients seen per year. The Department of Radiation Oncology at Sunnybrook is an academic unit fully affiliated with the University of Toronto. Palliative radiotherapy is one of the key research foci in the Department of Radiation Oncology. The Rapid Response Radiotherapy Program (RRRP) is a specialized clinic designed to provide timely palliative radiotherapy. The RRRP was developed in 1996, and approximately 500-600 patients are seen in one year. This program aims to improve the quality of life of palliative cancer patients, while decreasing wait time and allowing for same day treatment.

Contact

Professor Edward Chow, MBBS, PhD, FRCPC, Department of Radiation Oncology, Odette Cancer Centre, Sunnybrook Health Sciences Centre, 2075 Bayview Avenue, Toronto, Ontario, Canada M4N 3M5.
E-mail: Edward.Chow@sunnybrook.ca

In: Oncology
Editors: A. Furfari et al.
ISBN: 978-1-53610-608-4

Chapter 14

About the National Institute of Child Health and Human Development in Israel

The National Institute of Child Health and Human Development (NICHD) in Israel was established in 1998 as a virtual institute under the auspices of the Medical Director, Ministry of Social Affairs and Social Services in order to function as the research arm for the Office of the Medical Director. In 1998 the National Council for Child Health and Pediatrics, Ministry of Health and in 1999 the Director General and Deputy Director General of the Ministry of Health endorsed the establishment of the NICHD.

Mission

The mission of a National Institute for Child Health and Human Development in Israel is to provide an academic focal point for the scholarly interdisciplinary study of child life, health, public health, welfare, disability, rehabilitation, intellectual disability and related aspects of human development. This mission includes research, teaching, clinical work, information and public service activities in the field of child health and human development.

Service and academic activities

Over the years many activities became focused in the south of Israel due to collaboration with various professionals at the Faculty of Health Sciences (FOHS) at the Ben Gurion University of the Negev (BGU). Since 2000 an affiliation with the Zusman Child Development Center at the Pediatric Division of Soroka University Medical Center has resulted in collaboration around the establishment of the Down Syndrome Clinic at that center. In 2002 a full course on "Disability" was established at the Recanati School for Allied Professions in the Community, FOHS, BGU and in 2005 collaboration was started with the Primary Care Unit of the faculty and disability became part of the master of public health course on "Children and society". In the academic year 2005-2006 a one semester course on "Aging with disability" was started as part of the master of science program in gerontology in our collaboration with the Center for Multidisciplinary Research in Aging. In 2010 collaborations with the Division of Pediatrics, Hadassah Hebrew University Medical Center, Jerusalem, Israel around the National Down Syndrome Center and teaching students and residents about intellectual and developmental disabilities as part of their training at this campus.

Research activities

The affiliated staff have over the years published work from projects and research activities in this national and international collaboration. In the year 2000 the International Journal of Adolescent Medicine and Health and in 2005 the International Journal on Disability and Human Development of De Gruyter Publishing House (Berlin and New York) were affiliated with the National Institute of Child Health and Human Development. From 2008 also the International Journal of Child Health and Human Development (Nova Science, New York), the International Journal of Child and Adolescent Health (Nova Science) and the Journal of Pain Management (Nova Science) affiliated and from 2009 the International Public Health Journal (Nova Science) and Journal of Alternative Medicine Research (Nova Science). All peer-reviewed international journals.

National collaborations

Nationally the NICHD works in collaboration with the Faculty of Health Sciences, Ben Gurion University of the Negev; Department of Physical Therapy, Sackler School of Medicine, Tel Aviv University; Autism Center, Assaf HaRofeh Medical Center; National Rett and PKU Centers at Chaim Sheba Medical Center, Tel HaShomer; Department of Physiotherapy, Haifa University; Department of Education, Bar Ilan University, Ramat Gan, Faculty of Social Sciences and Health Sciences; College of Judea and Samaria in Ariel and in 2011 affiliation with Center for Pediatric Chronic Diseases and National Center for Down Syndrome, Department of Pediatrics, Hadassah Hebrew University Medical Center, Mount Scopus Campus, Jerusalem.

International collaborations

Internationally with the Department of Disability and Human Development, College of Applied Health Sciences, University of Illinois at Chicago; Strong Center for Developmental Disabilities, Golisano Children's Hospital at Strong, University of Rochester School of Medicine and Dentistry, New York; Centre on Intellectual Disabilities, University of Albany, New York; Centre for Chronic Disease Prevention and Control, Health Canada, Ottawa; Chandler Medical Center and Children's Hospital, Kentucky Children's Hospital, Section of Adolescent Medicine, University of Kentucky, Lexington; Chronic Disease Prevention and Control Research Center, Baylor College of Medicine, Houston, Texas; Division of Neuroscience, Department of Psychiatry, Columbia University, New York; Institute for the Study of Disadvantage and Disability, Atlanta; Center for Autism and Related Disorders, Department Psychiatry, Children's Hospital Boston, Boston; Department of Pediatric and Adolescent Medicine, Western Michigan University Homer Stryker MD School of Medicine, Kalamazoo, Michigan, United States; Department of Paediatrics, Child Health and Adolescent Medicine, Children's Hospital at Westmead, Westmead, Australia; International Centre for the Study of Occupational and Mental Health, Düsseldorf, Germany; Centre for Advanced Studies in Nursing, Department of General Practice and Primary Care, University of Aberdeen, Aberdeen, United Kingdom; Quality of Life Research Center, Copenhagen, Denmark; Nordic School of Public Health, Gottenburg, Sweden, Scandinavian Institute of Quality of Working Life, Oslo, Norway; The Department of Applied Social Sciences (APSS) of The Hong Kong Polytechnic University Hong Kong.

Targets

Our focus is on research, international collaborations, clinical work, teaching and policy in health, disability and human development and to establish the NICHD as a permanent institute in Israel in order to conduct model research and together with the four university schools of public health/medicine in Israel establish a national master and doctoral program in disability and human development at the institute to secure the next generation of professionals working in this often non-prestigious/low-status field of work.

Contact

Joav Merrick, MD, MMedSci, DMSc
Professor of Pediatrics
Medical Director, Health Services, Division for Intellectual and Developmental Disabilities, Ministry of Social Affairs and Social Services, POB 1260, IL-91012 Jerusalem, Israel.
E-mail: jmerrick@zahav.net.il

In: Oncology
Editors: A. Furfari et al.
ISBN: 978-1-53610-608-4

Chapter 15

ABOUT THE BOOK SERIES "HEALTH AND HUMAN DEVELOPMENT"

Health and human development is a book series with publications from a multidisciplinary group of researchers, practitioners and clinicians for an international professional forum interested in the broad spectrum of health and human development. Books already published:

- Merrick J, Omar HA, eds. Adolescent behavior research. International perspectives. New York: Nova Science, 2007.
- Kratky KW. Complementary medicine systems: Comparison and integration. New York: Nova Science, 2008.
- Schofield P, Merrick J, eds. Pain in children and youth. New York: Nova Science, 2009.
- Greydanus DE, Patel DR, Pratt HD, Calles Jr JL, eds. Behavioral pediatrics, 3 ed. New York: Nova Science, 2009.
- Ventegodt S, Merrick J, eds. Meaningful work: Research in quality of working life. New York: Nova Science, 2009.
- Omar HA, Greydanus DE, Patel DR, Merrick J, eds. Obesity and adolescence. A public health concern. New York: Nova Science, 2009.
- Lieberman A, Merrick J, eds. Poverty and children. A public health concern. New York: Nova Science, 2009.
- Goodbread J. Living on the edge. The mythical, spiritual and philosophical roots of social marginality. New York: Nova Science, 2009.
- Bennett DL, Towns S, Elliot E, Merrick J, eds. Challenges in adolescent health: An Australian perspective. New York: Nova Science, 2009.

- Schofield P, Merrick J, eds. Children and pain. New York: Nova Science, 2009.
- Sher L, Kandel I, Merrick J, eds. Alcohol-related cognitive disorders: Research and clinical perspectives. New York: Nova Science, 2009.
- Anyanwu EC. Advances in environmental health effects of toxigenic mold and mycotoxins. New York: Nova Science, 2009.
- Bell E, Merrick J, eds. Rural child health. International aspects. New York: Nova Science, 2009.
- Dubowitz H, Merrick J, eds. International aspects of child abuse and neglect. New York: Nova Science, 2010.
- Shahtahmasebi S, Berridge D. Conceptualizing behavior: A practical guide to data analysis. New York: Nova Science, 2010.
- Wernik U. Chance action and therapy. The playful way of changing. New York: Nova Science, 2010.
- Omar HA, Greydanus DE, Patel DR, Merrick J, eds. Adolescence and chronic illness. A public health concern. New York: Nova Science, 2010.
- Patel DR, Greydanus DE, Omar HA, Merrick J, eds. Adolescence and sports. New York: Nova Science, 2010.
- Shek DTL, Ma HK, Merrick J, eds. Positive youth development: Evaluation and future directions in a Chinese context. New York: Nova Science, 2010.
- Shek DTL, Ma HK, Merrick J, eds. Positive youth development: Implementation of a youth program in a Chinese context. New York: Nova Science, 2010.
- Omar HA, Greydanus DE, Tsitsika AK, Patel DR, Merrick J, eds. Pediatric and adolescent sexuality and gynecology: Principles for the primary care clinician. New York: Nova Science, 2010.
- Chow E, Merrick J, eds. Advanced cancer. Pain and quality of life. New York: Nova Science, 2010.
- Latzer Y, Merrick, J, Stein D, eds. Understanding eating disorders. Integrating culture, psychology and biology. New York: Nova Science, 2010.
- Sahgal A, Chow E, Merrick J, eds. Bone and brain metastases: Advances in research and treatment. New York: Nova Science, 2010.
- Postolache TT, Merrick J, eds. Environment, mood disorders and suicide. New York: Nova Science, 2010.
- Maharajh HD, Merrick J, eds. Social and cultural psychiatry experience from the Caribbean Region. New York: Nova Science, 2010.
- Mirsky J. Narratives and meanings of migration. New York: Nova Science, 2010.

- Harvey PW. Self-management and the health care consumer. New York: Nova Science, 2011.
- Ventegodt S, Merrick J. Sexology from a holistic point of view. New York: Nova Science, 2011.
- Ventegodt S, Merrick J. Principles of holistic psychiatry: A textbook on holistic medicine for mental disorders. New York: Nova Science, 2011.
- Greydanus DE, Calles Jr JL, Patel DR, Nazeer A, Merrick J, eds. Clinical aspects of psychopharmacology in childhood and adolescence. New York: Nova Science, 2011.
- Bell E, Seidel BM, Merrick J, eds. Climate change and rural child health. New York: Nova Science, 2011.
- Bell E, Zimitat C, Merrick J, eds. Rural medical education: Practical strategies. New York: Nova Science, 2011.
- Latzer Y, Tzischinsky. The dance of sleeping and eating among adolescents: Normal and pathological perspectives. New York: Nova Science, 2011.
- Deshmukh VD. The astonishing brain and holistic consciousness: Neuroscience and Vedanta perspectives. New York: Nova Science, 2011.
- Bell E, Westert GP, Merrick J, eds. Translational research for primary healthcare. New York: Nova Science, 2011.
- Shek DTL, Sun RCF, Merrick J, eds. Drug abuse in Hong Kong: Development and evaluation of a prevention program. New York: Nova Science, 2011.
- Ventegodt S, Hermansen TD, Merrick J. Human Development: Biology from a holistic point of view. New York: Nova Science, 2011.
- Ventegodt S, Merrick J. Our search for meaning in life. New York: Nova Science, 2011.
- Caron RM, Merrick J, eds. Building community capacity: Minority and immigrant populations. New York: Nova Science, 2012.
- Klein H, Merrick J, eds. Human immunodeficiency virus (HIV) research: Social science aspects. New York: Nova Science, 2012.
- Lutzker JR, Merrick J, eds. Applied public health: Examining multifaceted Social or ecological problems and child maltreatment. New York: Nova Science, 2012.
- Chemtob D, Merrick J, eds. AIDS and tuberculosis: Public health aspects. New York: Nova Science, 2012.
- Ventegodt S, Merrick J. Textbook on evidence-based holistic mind-body medicine: Basic principles of healing in traditional Hippocratic medicine. New York: Nova Science, 2012.
- Ventegodt S, Merrick J. Textbook on evidence-based holistic mind-body medicine: Holistic practice of traditional Hippocratic medicine. New York: Nova Science, 2012.

- Ventegodt S, Merrick J. Textbook on evidence-based holistic mind-body medicine: Healing the mind in traditional Hippocratic medicine. New York: Nova Science, 2012.
- Ventegodt S, Merrick J. Textbook on evidence-based holistic mind-body medicine: Sexology and traditional Hippocratic medicine. New York: Nova Science, 2012.
- Ventegodt S, Merrick J. Textbook on evidence-based holistic mind-body medicine: Research, philosophy, economy and politics of traditional Hippocratic medicine. New York: Nova Science, 2012.
- Caron RM, Merrick J, eds. Building community capacity: Skills and principles. New York: Nova Science, 2012.
- Lemal M, Merrick J, eds. Health risk communication. New York: Nova Science, 2012.
- Ventegodt S, Merrick J. Textbook on evidence-based holistic mind-body medicine: Basic philosophy and ethics of traditional Hippocratic medicine. New York: Nova Science, 2013.
- Caron RM, Merrick J, eds. Building community capacity: Case examples from around the world. New York: Nova Science, 2013.
- Steele RE. Managed care in a public setting. New York: Nova Science, 2013.
- Srabstein JC, Merrick J, eds. Bullying: A public health concern. New York: Nova Science, 2013.
- Pulenzas N, Lechner B, Thavarajah N, Chow E, Merrick J, eds. Advanced cancer: Managing symptoms and quality of life. New York: Nova Science, 2013.
- Stein D, Latzer Y, eds. Treatment and recovery of eating disorders. New York: Nova Science, 2013.
- Sun J, Buys N, Merrick J. Health promotion: Community singing as a vehicle to promote health. New York: Nova Science, 2013.
- Pulenzas N, Lechner B, Thavarajah N, Chow E, Merrick J, eds. Advanced cancer: Managing symptoms and quality of life. New York: Nova Science, 2013.
- Sun J, Buys N, Merrick J. Health promotion: Strengthening positive health and preventing disease. New York: Nova Science, 2013.
- Merrick J, Israeli S, eds. Food, nutrition and eating behavior. New York: Nova Science, 2013.
- Shahtahmasebi S, Merrick J. Suicide from a public health perspective. New York: Nova Science, 2014.
- Merrick J, Tenenbaum A, eds. Public health concern: Smoking, alcohol and substance use. New York: Nova Science, 2014.

- Merrick J, Aspler S, Morad M, eds. Mental health from an international perspective. New York: Nova Science, 2014.
- Merrick J, ed. India: Health and human development aspects. New York: Nova Science, 2014.
- Caron R, Merrick J, eds. Public health: Improving health via inter-professional collaborations. New York: Nova Science, 2014.
- Merrick J, ed. Pain Mangement Yearbook 2014. New York: Nova Science, 2015.
- Merrick J, ed. Public Health Yearbook 2014. New York: Nova Science, 2015.
- Sher L, Merrick J, eds. Forensic psychiatry: A public health perspective. New York: Nova Science, 2015.
- Shek DTL, Wu FKY, Merrick J, eds. Leadership and service learning education: Holistic development for Chinese university students. New York: Nova Science, 2015.
- Calles JL, Greydanus DE, Merrick J, eds. Mental and holistic health: Some international perspectives. New York: Nova Science, 2015.
- Lechner B, Chow R, Pulenzas N, Popovic M, Zhang N, Zhang X, Chow E, Merrick J, eds. Cancer: Treatment, decision making and quality of life. New York: Nova Science, 2016.
- Lechner B, Chow R, Pulenzas N, Popovic M, Zhang N, Zhang X, Chow E, Merrick J, eds. Cancer: Pain and symptom management. New York: Nova Science, 2016.
- Lechner B, Chow R, Pulenzas N, Popovic M, Zhang N, Zhang X, Chow E, Merrick J, eds. Cancer: Bone metastases, CNS metastases and pathological fractures. New York: Nova Science, 2016.
- Lechner B, Chow R, Pulenzas N, Popovic M, Zhang N, Zhang X, Chow E, Merrick J, eds. Cancer: Spinal cord, lung, breast, cervical, prostate, head and neck cancer. New York: Nova Science, 2016.
- Lechner B, Chow R, Pulenzas N, Popovic M, Zhang N, Zhang X, Chow E, Merrick J, eds. Cancer: Survival, quality of life and ethical implications. New York: Nova Science, 2016.
- Davidovitch N, Gross Z, Ribakov Y, Slobodianiuk A, eds. Quality, mobility and globalization in the higher education system: A comparative look at the challenges of academic teaching. New York: Nova Science, 2016.
- Henry B, Agarwal A, Chow E, Omar HA, Merrick J, eds. Cannabis: Medical aspects. New York: Nova Science, 2017.

Contact

Professor Joav Merrick, MD, MMedSci, DMSc
Medical Director, Health Services
Division for Intellectual and Developmental Disabilities
Ministry of Social Affairs and Social Services
POBox 1260, IL-91012 Jerusalem, Israel
E-mail: jmerrick@zahav.net.il

SECTION FOUR: INDEX

INDEX

#

A

B

C

D

E

F

G

H

I

N

O

P

Q

R

S

T

U

V

W

Y

Z